Space Competition and the Dynamics of Conflict

Using Game Theory and Artificial Intelligence to Gain Strategic Insight

BONNIE TRIEZENBERG, KRISTA LANGELAND, BRYCE DOWNING

Prepared for the National Reconnaissance Office
Approved for public release; distribution unlimited

RAND NATIONAL DEFENSE RESEARCH INSTITUTE

For more information on this publication, visit **www.rand.org/t/RRA751-1**.

Published by the RAND Corporation, Santa Monica, Calif.

Library of Congress Cataloging-in-Publication Data is available for this publication.
ISBN: 978-1-9774-0809-9

About This Report

The authors of this report summarize research to date about using a game-theoretic model of competition and cooperation in outer space to examine the dynamics of conflict there, short of wars of total destruction. In 2014, the RAND Corporation began developing a game-theoretic model to assess strategic implications of investments in space capabilities by the United States and a competitor nation. In projects since, we built on traditional game theory to provide a context-rich assessment of how nation-state investments may play out over a range of possible futures. Although prior published research using this model explored the impact of investments on deterring horizontal escalation of a terrestrial war into outer space, this report focuses on the *dynamics of space competition* in the context of limited space war. This report provides a summary of the aspects of our work that should be of interest to a general audience of space policymakers and potential warfighters.

The research reported here was completed in March 2021 and underwent security review with the sponsor and the Defense Office of Prepublication and Security Review before public release.

RAND National Security Research Division

This research is sponsored by the National Reconnaissance Office (NRO) and conducted within the Cyber and Intelligence Policy Center of the RAND National Security Research Division (NSRD), which operates the RAND National Defense Research Institute (NDRI), a federally funded research and development center (FFRDC) sponsored by the Office of the Secretary of Defense, the Joint Staff, the Unified Combatant Commands, the Navy, the Marine Corps, the defense agencies, and the defense intelligence enterprise.

For more information on the Cyber and Intelligence Policy Center, see www.rand.org/nsrd/intel or contact the director (contact information is provided on the webpage).

Acknowledgments

The analysis and discussion in this report would not have been possible without a broad spectrum of contributors. We are truly grateful for all of the comments, insights, and engaging discussions that have ensued as a result of this research, and we wish to extend our thanks for these numerous contributions to the analysis discussed in this report.

We thank Stewart M. Cameron, who first posed the original problem of folding prospect theory into game theory and provided us with prescient insights throughout the project, which contributed to our work. We are grateful to Geoffrey Torrington and Dave Baiocchi, who developed this idea in its nascent stage and provided key guidance on the mission needs of the intelligence community. This analysis would not have happened without the close collaboration of James Pita, who not only wrote the code for the initial version of the game but also provided game-theory expertise throughout our many uses of and upgrades to the game-theoretic model.

We are grateful to all of our previous sponsors at the National Reconnaissance Office's Survivability Assurance Office, which helped us along the way. Robert Lee opened many doors and increased the visibility of our work across the national security space community, and Mark Williams and Scott V provided invaluable insight and guidance to ensure that our work remained relevant to the specific concerns of the intelligence community. Commentary by Lt Gen John Shaw and Lt Gen Stephen Whiting of the (now) U.S. Space Force encouraged us to dig deeper into the dynamics of conflict.

Many RAND colleagues provided support in innumerable ways throughout the past five years. We are indebted to numerous colleagues, including Forrest Morgan, Bruce McClintock, and Steven Flanagan, for their insights into strategic messaging and deterrence in the space domain; Karen Lee, Alex Sedlack, and Ben Goirigolzarri contributed greatly to the analysis. The dissertation committees who supported both Bonnie and Ben's use of the model also deserve our thanks—Scott Pace (then at George Washington University) and William Shelton, Jonathan Wong, and Aaron Frank. Finally, we thank our colleagues Bruce McClintock and Jonathan Welburn for their thoughtful review of this report. As always, however, we retain full authorship of this work and any deficiencies are ours alone.

Summary

As space becomes increasingly militarized, understanding the long-term impact of messaging, investments, and actions becomes increasingly vital to maintaining continued use of space.[1] Starting in 2014, the RAND Corporation developed a game-theoretic model with the objective of providing the National Reconnaissance Office's Survivability Assurance Office with modeling capability to assess future conflicts in space and the implications for investments in new space capabilities. The model has since been used in a variety of studies, both to inform U.S. government investment decisions regarding space and as the basis for two doctoral dissertations examining the nature of space warfare. Although prior publications using this model have broadly discussed factors that influence U.S. ability to deter the horizontal escalation of terrestrial war into the space domain,[2] this report focuses less on those issues and instead examines the dynamics of conflict, short of wars of total destruction, to discuss specific questions of interest to U.S. space policymakers.

The questions and a brief summary of our findings are as follows:

- Assuming space domain awareness (SDA) improves a nation's ability to defend against kinetic attacks, can those improved defenses deter the adversary from attacking? Do they change conflict outcomes?

Our work examining SDA-enhanced defenses revealed a clear principle of conflict dynamics in space or in any domain where there are multiple

---

[1] Increased public profile of militarized space instigated this research. Although space historically has been an area where national competition plays out and has had military uses, its most frequent uses are commercial and civil. An examination of the current space catalog shows that less than 30 percent of the satellites in orbit have been designed for military use. See Bonnie L. Triezenberg, *Deterring Space War: An Exploratory Analysis Incorporating Prospect Theory into a Game Theoretic Model of Space Warfare*, Santa Monica, Calif.: RAND Corporation, RGSD-400, 2017.

[2] See, for example, Benjamin Goirigolzarri, *A Need for Speed? Identifying the Effects of Space Acquisition Timelines on Space Deterrence and Conflict Outcomes*, Santa Monica, Calif.: RAND Corporation, RGSD-432, 2019, and Triezenberg, 2017.

attack vectors: improving defenses against one type of attack (in this case, kinetic) simply shifts the adversary's tactics, not their strategy. If it is in the adversary's long-term strategic interest to attack space assets,[3] reducing the probability of success for their preferred attack option may simply mean they will switch to the next best weapon-target pairing. Although we did not find that SDA-enhanced defenses gave us better deterrence or better conflict outcomes, it did clearly push the adversary away from kinetic attacks and toward less debris-generating attacks—a result that enhances society's ability to preserve the use of space for future generations.

- Are misperceptions regarding an opponent's offensive capabilities (i.e., their ability to hold space assets at risk) stabilizing or destabilizing? Do these misperceptions change conflict outcomes?

In our examination of conflict between near-peer competitors, the effect of offensive misperceptions is quite nuanced. We found misperception to be stabilizing when (1) the adversary perceives a peer opponent to be disadvantaged or (2) a disadvantaged opponent is perceived to be a peer.[4] In other conditions, we found misperception to be destabilizing or to have no clear trend. Decisions on whether to reveal or conceal investments in, or the actual extent of, offensive capabilities are not straightforward, especially when considering that not all potential opponents in space are peer competitors.

- When conflict in space occurs, do the nations engage in distinct attack strategies similar to those observed in games of chess or other strategic competitions? Related questions include the following:

[3] A *space asset* need not be located in space. Space assets include terrestrial monitoring and control stations, user terminals, sensing systems, the links that connect them, and the humans that use them. Similarly, not all space weapons are placed in space. In fact, most space weapons today are located either in cyber space or in terrestrial locations (e.g., ground-based missiles, laser weapons, jammers).

[4] We say that an effect is *stabilizing* if the resultant conflict is of lower intensity. Our metric for conflict intensity, described in Chapter Two, takes into account the perceived escalatory nature of attacks as well as the character and number of attacks.

a. Can we identify the conditions that give rise to specific attack strategies?
b. Can investments in specific offensive or defensive capabilities be used to shape an opponent's attack strategy?
c. Can revelation or concealment of investments in offensive or defensive capabilities be used to shape an opponent's attack strategy?

We found three distinct patterns of strategic interaction (i.e., attack strategies) in games in which deterrence was lost but that stopped short of wars of total destruction. We term the first *early shaping of the battlefield.* In these games, attacks occurred two or more years before ground conflict. We observed two possible motivations for attackers to adopt this strategy: (1) as an attempt to shift the opponent away from investing in weapons and into investing in recovery of the space capabilities lost in the attack or (2) as a response to the opponent having created a situation in which a player needs to use a weapon while it is still effective (i.e., the opponent has created a use-it-or-lose-it scenario).[5]

We also observed a cluster of games that are characterized as *wars of weapons attrition.* In these games, both sides reserve their attacks until just before or simultaneous with the start of a ground war and the attacks from both sides are focused primarily on reducing their opponent's offensive capability.

Finally, there is a set of games where we observed a classic *horizontal escalation* of the ground war into the space domain. In these games, attacks are made well after the start of the terrestrial conflict and focus on creating exploitable shortfalls in the adversary's ability to project power from space.

In this report, we offer only tentative observations regarding the conditions that give rise to the discussed strategic interaction patterns and possible ways a nation might shape an opponent's attack strategies. Although

[5] We also found a pattern of early shaping of the battlefield that we term *brinksmanship.* It is unclear whether this pattern is simply a manifestation of known issues with game-theoretic models or if it represents a pattern we might expect to find in the real world. The brinksmanship pattern is found in many of our games in which both sides engage in a war of total destruction. See Appendix C for a detailed description of the brinksmanship pattern.

we can re-create the logic for any individual decision made by the players in our games, clear correlations that might lead us to causal theories for why a particular strategy emerges are less readily observable.

In fact, our experience in using complex artificial intelligence (AI) to explore strategy leads us to offer three observations about that use:

- We have come to believe that a complex game-theoretic model is incapable of revealing anything beyond correlation and, even then, only if we have a theoretical basis for our inquiry. A hypothesis can then be used to construct counterfactual game scenarios that can be run to test whether A does in fact lead to B.[6] More often than not, however, we found that factors that we hypothesized might be causal had no impact on overall strategies. There are simply too many substitutes available in the game for any one factor to be causal.[7]
- We find ample evidence of the butterfly effect of chaos theory in our work. Very small differences in initial conditions or individual decisions, which we might term *negligible* in the short term, often have profound effects in the longer term. As Edward Lorenz states in his seminal talk on the butterfly effect, the best model for research may not be the most exact model of the world.[8] Simpler models may be more useful.[9]

---

[6] This approach is called *causal AI*. For a top-level view, see Judea Pearl and Dana Mackenzie, *The Book of Why: The New Science of Cause and Effect*, New York: Basic Books, 2018.

[7] Where we could construct and find evidence to support specific theories, that support depended on *combinations* of factors. However, we were unsuccessful in our attempts to use pattern recognition algorithms, such as those used in machine learning, to detect combinations of factors that would be explanatory of strategies. We suspect this is because we did not have adequate instrumentation to capture emergent properties over the course of the game.

[8] The text of Lorenz's December 1972 speech can be found in Ralph Abraham and Yoshisuke Ueda, and Yoshisuke Ueda, eds., *The Chaos Avant-Garde: Memoirs of the Early Days of Chaos Theory*, World Scientific Series on Nonlinear Science, Series A, Vol. 39, Singapore: World Scientific, 2000.

[9] Branislav L. Slantchev argues the particular case that simpler game-theoretic models are most useful in the study of conflict. However, as he notes, the decision of what is "simple enough" is not straightforward (see Branislav L. Slantchev, "On the Proper Use

- None of the research conducted with the model proceeded in a linear fashion. Runs with our initial models inevitably "taught" us some key aspect of the problem. We then "taught" the model to incorporate that learning, sometimes thru multiple iterations, with the net result being what we term incremental human-AI collaborative learning.[10]

Recall that the primary goal of our multiyear effort is to assess strategic implications of investment decisions in space capabilities by the United States and a competitor nation. In our prior publications, we expounded on the limitations of the model and highlighted that a game-theoretic model is only capable of exploratory research.[11] Expanding the research goals to specific investments and to questions of whether to hide or reveal those investments has highlighted those limitations—as we just noted, although we can distinguish clear patterns of strategic interactions, the factors that give rise to them often remain shrouded in complexity. This does not mean, however, that our study of these questions and the dynamics of conflict are of no value. We believe the value of the model is that it expands our imagination about the range of outcomes that are possible when contemplating how, when, and whether future conflicts may horizontally escalate into the space domain. It is this range of possible outcomes that we strive to communicate

of Game-Theoretic Models in Conflict Studies," *Peace Economics, Peace Science and Public Policy*, Vol. 23, No. 4, December 14, 2017). We discuss some of our more notable failures in modeling appropriate abstractions in Appendix C, in the hope that others may benefit from our experience.

[10] A reasonable body of applicable research in human-AI collaborative learning comes from observations about how the game of chess has changed since the first computers became smart enough to defeat the leading human strategists. For a top-level view, see Dirk Knemeyer and Jonathan Follet, "How 22 Years of AI Superiority Changed Chess," *Towards Data Science*, March 5, 2019. A more academic treatise on the subject of human-AI collaborative learning can be found in Nan-ning Zheng, Zi-yu Liu, Peng-ju Ren, Yong-qiang Ma, Shi-tao Chen, Si-yu Yu, Jian-ru Xue, Ba-dong Chen, and Fei-yue Wang, "Hybrid-Augmented Intelligence: Collaboration and Cognition," *Frontiers of Information Technology and Electronic Engineering*, Vol. 18, No. 2, 2017.

[11] Triezenberg, 2017.

in this report and that we believe will benefit future space warfighters and strategists.[12]

---

[12] We should also note what we are not trying to communicate in this report. First, we do not provide an in-depth discussion of the threats against space assets in this report. For the past three years, both the Secure World Foundation and the Center for Strategic and International Studies have issued excellent unclassified summaries on that topic and our game model is capable of modelling the full range of threats mentioned in those summaries. See Brian Weedon and Victoria Samson, *Global Counterspace Capabilities: An Open Source Assessment*, Broomfield, Colo.: Secure World Foundation, April 2020, and Todd Harrison, Kaitlyn Johnson, Thomas G. Roberts, Tyler Way, and Makena Young, *Space Threat Assessment 2020: A Report of the CSIS Aerospace Security Project*, Washington, D.C.: Center for Strategic and International Studies, 2020. Second, we do not provide an in-depth discussion of the physics that determine the effectivity of specific weapons/target pairing or of specific defenses against those weapons. In our model, we incorporate the effectivities of weapons/target/defense as a function of attack intent (deny, deceive, disrupt, degrade, destroy), as well as the attributes of the different weapons, targets, and defenses. We do not, however, model the underlying physics. A top-level, but reasonably thorough, review of those physics is the topic of a recently released report by the Aerospace Corporation, and we refer the interested reader to that report: Rebecca Reesman and James R. Wilson, *The Physics of Space War: How Orbital Dynamics Constrain Space to Space Engagements*, El Segundo, Calif.: Aerospace Corporation, October 2020. Third, the purpose of this report is not to provide the detail necessary to reproduce our model or our results. Its only purpose is to pass on the lessons we have learned about the nature of space competition and the nature of research when using a complex AI.

Contents

Figures and Tables

Figures

Tables

CHAPTER ONE

Introduction

Purpose of This Report

This report summarizes research to date using a game-theoretic model of competition and cooperation in outer space to examine the dynamics of conflict in outer space, short of wars of total destruction. In 2014, the RAND Corporation began developing a game-theoretic model to assess strategic implications of investments in space capabilities by the United States and a competitor nation. Projects since have built on traditional game theory to provide a context-rich assessment of how nation-state investments may play out over a range of possible futures. Although prior published research using this model explored the effect of investments on deterring horizontal escalation of a terrestrial war into outer space, this report focuses on the *dynamics of space competition*. In the context of limited space war, this report focuses on those dynamics while exploring specific questions of interest to U.S. space policymakers. These include the following:

1. Assuming space domain awareness (SDA) improves a nation's ability to defend against kinetic attacks, can those improved defenses deter the adversary from attacking? Do they change conflict outcomes?
2. Are misperceptions regarding an opponent's offensive capabilities (i.e., their ability to hold space assets at risk) stabilizing or destabilizing? Do these misperceptions change conflict outcomes?
3. When conflict in space occurs, do the nations engage in distinct attack strategies similar to those observed in games of chess or other strategic competitions? Related questions include the following:
    a. Can we identify the conditions that give rise to specific attack strategies?

b. Can investments in specific offensive or defensive capabilities be used to shape an opponent's attack strategy?
c. Can revelation or concealment of investments in offensive or defensive capabilities be used to shape an opponent's attack strategy?

Our sponsoring office supported dissemination of results of our analyses concerning these questions throughout 2018 and 2019 to the U.S. intelligence community and U.S. Space Command. The impetus for this unclassified report, therefore, is not to duplicate those reports and briefings but to provide a summary of those aspects of our work that would be of interest to a more general audience of space policymakers and potential warfighters.

In addition to discussing these specific questions, this report references prior studies in which the model was used[1] and includes general observations from across that body of research regarding the nature of deterrence and conflict in outer space. The primary goal of our multiyear effort is to assess strategic implications of investment decisions in space capabilities by the United States and a competitor nation. In our previous publications, we expounded on the limitations of the model and highlighted that a game-theoretic model is only capable of exploratory research.[2] Expanding the research goals to specific investments and to questions of whether to hide

[1] Previous unclassified studies focused on the ability to deter horizontal escalation of a terrestrial war into the space domain. The most recent examines the impact on deterrence and space superiority of asymmetries in acquisition speed for space systems (Benjamin Goirigolzarri, *A Need for Speed? Identifying the Effects of Space Acquisition Timelines on Space Deterrence and Conflict Outcomes*, Santa Monica, Calif.: RAND Corporation, RGSD-432, 2019). Earlier work explores how the offensive/defensive balance and a nation's asymmetric dependence on space to project power impacts deterrence (Bonnie Triezenberg, *Deterring Space War: An Exploratory Analysis Incorporating Prospect Theory into a Game Theoretic Model of Space Warfare*, Santa Monica, Calif.: RAND Corporation, RGSD-400, 2017). This earlier work also provides the most comprehensive publicly accessible overview of the game model. A more complete description of the game and details regarding model validation are contained in Geoffrey Torrington, Bonnie L. Triezenberg, Krista Langeland, Lisa Saum-Manning, Timothy Marler, Elizabeth M. Bartels, and James Pita, *Exploring Space Deterrence: Final Phase II Report: Using Game Theory and Prospect Theory to Inform Future Strategies*, Santa Monica, Calif.: RAND Corporation, 2019, Not available to the general public.

[2] Triezenberg, 2017.

or reveal those investments has highlighted those limitations: Although we can distinguish clear patterns of strategic interactions, the factors that give rise to them often remain shrouded in complexity. This does not mean however, that our study of these questions and the dynamics of conflict are of no value. We believe the value of the model is that it expands our imagination to the range of outcomes that are possible when contemplating how, when, and whether future conflicts may horizontally escalate into the space domain. It is this range of possible outcomes that we strive to communicate in this report and that we believe will benefit future space warfighters and strategists.[3]

Space as a Warfighting Domain

Since the first Sputnik launched in 1957, outer space has been a place where competition between great national powers has played out. We have pointed

[3] We should also note what we are *not* trying to communicate in this report. First, we do not provide an in-depth discussion of the threats against space assets. For the past three years, both the Secure World Foundation and the Center for Strategic and International Studies have issued excellent unclassified summaries on that topic, and our game model is capable of modeling the full range of threats mentioned in those summaries. See Brian Weedon and Victoria Samson, *Global Counterspace Capabilities: An Open Source Assessment*, Broomfield, Colo.: Secure World Foundation, April 2020, and Todd Harrison, Kaitlyn Johnson, Thomas G. Roberts, Tyler Way, and Makena Young, *Space Threat Assessment 2020: A Report of the CSIS Aerospace Security Project*, Washington, D.C.: Center for Strategic and International Studies, 2020. Second, we do not provide an in-depth discussion of the physics that determine the effectivity of specific weapons/target pairing or of specific defenses against those weapons. In our model, we incorporate the effectivities of weapons/target/defense as a function of attack intent (deny, deceive, disrupt, degrade, destroy), as well as the attributes of the different weapons, targets, and defenses. We do not, however, model the underlying physics. A top-level, but reasonably thorough, review of those physics is the topic of a recently released report by the Aerospace Corporation, and we refer the interested reader to that report: Rebecca Reesman and James R. Wilson, *The Physics of Space War: How Orbital Dynamics Constrain Space to Space Engagements*, El Segundo, Calif.: Aerospace Corporation, October 2020. Third, the purpose of this report is not to provide the detail necessary to reproduce our model or our results. Its only purpose is to pass on the lessons we have learned about the nature of space competition and the nature of research when using a complex artificial intelligence (AI).

out in other work that today's reemergent discussion of space as a war fighting domain may be the third such wave.[4] However, this is the first time in history that these sentiments have prompted today's great-power competitors to openly reorganize their respective militaries to prepare to conduct warfighting operations in space. As part of a major military reorganization, China created the People's Liberation Army (PLA) Strategic Support Force (SSF) in 2015. The SSF is tasked with the development and employment of the PLA's space capabilities and indicates a shift toward PLA prioritization of space.[5] Also in 2015, Russia merged several branches of its military to form the Aerospace Forces. Combining Russia's air force, air defense, anti-missile, and space forces, this reorganization is a recognition of the role of space in modern warfare and the increasing importance of integration with the other branches.[6] And, in 2019, the United States both reestablished outer space as a geographic warfighting command (U.S. Space Command)[7] and directed the establishment of the U.S. Space Force (USSF) as the sixth branch of its armed forces, recognizing this as a "strategic imperative" in recognition of the growing role of space in national and economic security.[8]

Despite the long-term competition in space and these recent reorganizations, strategic theories for space warfighting are largely underdevel-

---

4 The first wave occurred in the early 1960s, when both the United States and Russia openly experimented with anti-satellite weapons. The second wave occurred in the mid-1980s and included a U.S. demonstration of a kinetic kill of a satellite in low-earth orbit. For a discussion of these waves and the historic context in which they emerged and receded, see Triezenberg, 2017.

5 See a discussion on the creation of the SSF and the related implications of the PLA reorganization in Kevin L. Pollpeter, Michael S. Chase, and Eric Heginbotham, *The Creation of the PLA Strategic Space Force and Its Implications for Chinese Military Space Operations*, Santa Monica, Calif.: RAND Corporation, RR-2058-AF, 2017.

6 See, i.e., Matthew Bodner, "Russian Military Merges Air Force and Space Command," *Moscow Times*, August 3, 2015.

7 U.S. Space Command was originally created in 1985 but was inactivated in 2002 when its responsibilities were merged into the U.S. Strategic Command. It was reestablished in August 2019.

8 Mark Esper, quoted in U.S. Department of Defense, "Department of Defense Establishes U.S. Space Force," news release, December 20, 2019.

oped and unexplored.[9] A motivating factor in the recent establishment of the USSF is the desire to develop strategic theories of space warfare distinct from theories of land, naval, and air warfare in anticipation of warfighting operations. The space domain is unique among physical domains because of its global presence provided by "combining the high-altitude perspective of space with the enduring longevity of forward employed spacecraft and an international legal regime which recognizes overflight of any point on the Earth by spacecraft. This affords unique opportunities for military power."[10] There is a growing recognition that space is its own warfighting domain. GEN Mark Milley, Chairman of the Joint Chiefs of Staff, asserts that "[i]n military operations, space is not just a place from which we support combat operations in other domains, but a warfighting domain in and of itself."[11] Indeed, there are legitimate reasons to believe that space warfare *will* give rise to significantly different strategies than those seen in land, naval, and air warfare. After all, both the weapons and targets of space warfare will be largely unmanned, there is no national "homeland" in space, attacks will be largely invisible to ordinary citizens (although their impact may be highly visible), and, with a few notable exception, the tactics, training, and procedures (TTPs) needed to successfully execute or defend against attacks in space require highly specialized technical knowledge.[12]

[9] We note a growing number of colleagues who are working to fill this gap. For additional reading see Bleddyn E. Bowen, *War in Space: Strategy, Spacepower, Geopolitics*, Edinburgh: Edinburgh University Press, 2020; John J. Klein, *Space Warfare: Strategy, Principles and Policy*, New York: Routledge, 2006; and Howard Kleinberg, "On War in Space," *Astropolitics: The International Journal of Space Politics and Policy*, Vol. 5, No. 1, 2007.

[10] U.S. Space Force, *Space Capstone Publication: Spacepower, Doctrine for Space Forces*, Arlington, Va.: Headquarters, U.S. Space Force, June 2020, p. 22.

[11] U.S. Department of Defense, 2019.

[12] Much of that specialization is because of the different physics of space. For instance, the Army, Navy, and Air Force can all assume that, if they fire a bullet or other free-flying projectile, it will follow a ballistic trajectory. This is not true in space. Although a ballistic trajectory is something anyone who has thrown a rock or a baseball has experienced, space physics are dominated by centripetal forces that produce orbital trajectories. The motion that results from firing a projectile in space is one that few humans have developed an intuition for. See Reesman and Wilson, 2020, for a more in-depth examination of how the physics of space may impact warfighting strategies and tactics.

In this report, we discuss both strategy and tactics in space warfare and the effects of investment decisions on both. *Strategy*, as used in this report specifically and in game theory more generally, is the overall approach to cooperation and conflict that a nation adopts after fully considering how its actions will be perceived by allies and adversaries, how those perceptions might influence subsequent actions by those parties, and whether the long-term outcomes are likely to support core national interests.[13] Thus, it requires a keen understanding of the competitive environment, including other nations' goals, objectives, and capabilities and the ability to anticipate and shape perceptions. *Tactics* are the means by which strategy is carried out.[14] As a concrete example, we classify the decision to execute an attack to degrade an adversary's ability to project military power from space as a *strategic decision*. Our use of this term implies an assumption that the decision was made after considering the ramifications of the attack and ascertaining that there is a high probability that the attack will result in conditions favorable to national objectives. However, we classify the decisions regarding which weapon/target pairing to use when executing the attack as tactical, as long as that pairing achieves the desired degradation within the constraints of the strategic goal.[15]

Developing strategy is hard. Decisionmaking in a competitive environment inherently requires consideration of numerous factors beyond your own net objectives and, possibly more importantly, beyond your control. The adversary is also trying to meet their own objectives, which might not

---

[13] Our definition of *strategy* is similar to that used by Joint Publication 3-0, which defines *strategy* as "A prudent idea or set of ideas for employing the instruments of national power in a synchronized and integrated fashion to achieve theater, national, and/or multinational objectives." (Joint Publication 3-0, *Joint Operations*, Washington, D.C.: Joint Chiefs of Staff, January 17, 2017, incorporating change 1, October 22, 2018, p. GL-15).

[14] For *tactics*, we use Merriam-Webster's definition: "the art or skill of employing available means to accomplish an end." We find this preferable to military definitions, which emphasize the employment of forces. Our games often demonstrate how the simple ability to hold an opponent's assets or weapons at risk (rather than actual employment of force) is the most effective tactic in implementing a player's strategy.

[15] This is not to say that weapon/target pairing is never a strategic decision. The legitimacy of both targets and the weapons used against them can heavily influence the ability to meet strategic goals.

be the objectives you assume they have. Furthermore, your own actions may cause the adversary to shift their objectives—conflict is dynamic.[16] Understanding *perceptions* in the short and long term becomes a key aspect of successful strategic decisionmaking. In the heat of conflict, this craft relies heavily on intuition developed over years of observations of action and reaction; there is rarely time to think through thousands of possible scenarios about how decisions may play out. Luckily, tools are available that allow us to build and sharpen strategic thinking skills and to build intuition prior to conflict.[17] This report describes one such tool—a game-theoretic model that allows us to explore long-term impacts of decisionmaking about the space domain in a future conflict environment. It also describes strategic principles regarding conflict in space that we have come to appreciate through our years of working with this game-theoretic model and related research. Finally, it describes common patterns of strategic interactions between competitors in space that we observed in our research.[18]

[16] Although volumes have been written on strategy in international relations (several of which are included in the references), from our own reading, we recommend Axelrod's *The Evolution of Cooperation* both because it concisely and plainly explains difficult concepts and because it focuses on the dynamics of conflict—the subject of this report (Robert Axelrod, *The Evolution of Cooperation*, New York: Basic Books, 1984).

[17] Both strategic and expert intuition are needed in conflict. As William Duggan says in the introduction of *Strategic Intuition, the Creative Spark in Human Achievement* (New York: Columbia Business School Publishing, [2007] 2013),

> Ordinary intuition is just a feeling, a gut instinct. Expert intuition is snap judgments, when you instantly recognize something familiar, the way a tennis pro knows where the ball will go from the arc and speed of the opponent's racket. . . . The third kind, strategic intuition, is not a vague feeling, like ordinary intuition. Strategic intuition is a clear thought. And it's not fast, like expert intuition. It's slow. That flash of insight you had last night might solve a problem that's been on your mind for a month. And it doesn't happen in familiar situations, like a tennis match. Strategic intuition works in new situations. That's when you need it most.

[18] Because of the classified nature of much of our work, the description of the model itself and its inputs and outputs are discussed at only a summary level in this report.

Strategic Competition and Game Theory

As noted earlier, game-theoretic models are one way to develop the strategic intuition we need regarding conflict and cooperation in space. This is because game theory applies a model of "conflict and cooperation between intelligent rational decision-makers" to reason about how war might play out under a theory of *strategic competition*.[19] Strategic competition contrasts with a theory of perfect competition, which assumes that there are numerous participants, none of whom can exert outsized influence on the other participants (i.e., no individual player actions influence the equilibrium outcome).[20] Theories of strategic competition, on the other hand, explicitly model that moves of one player impact the outcome and payoffs for the other in solving for the equilibrium outcomes. Strategic competition is studied in poker, chess, go, and many other games of strategy, as well as in real-life economic, political, and social interaction.[21] Although poker, go, and chess all have features that are echoed in a strategic competition in space, we identify the game of chess as the most analogous for our work. In chess, the value of an action depends significantly on context and changes over the timespan of game play. In a strategic competition in space, we find that the value of any action is heavily time- and context-dependent. Furthermore, this context is shaped by prior player actions. The value of a particular action or investment in a strategic game in space, just as in chess, can shift wildly with a different set of opening moves or opposing player strategies. Finally, success in both chess and space strategy depend on the capability to hold an opponent's resources at risk. This is a significant difference from competitions centered on gaining and holding territory.

Game-theoretic models can be an effective means to study strategic competition because they are structured as a simultaneous optimization problem—as researchers, we "solve" the model for the set of actions that

[19] Roger B. Myerson, *Game Theory: Analysis of Conflict*, Cambridge, Mass.: Harvard University Press, 1991, p. 1.

[20] Gerard Debreu, *Theory of Value: An Axiomatic Analysis of Economic Equilibrium*, New Haven, Conn.: Yale University Press, 1959.

[21] Game theory came to prominence in the Cold War to explain how the strategic balance of power could be used to deter nuclear war.

result in the optimal achievement of each party's goals *subject to* the opposing party also achieving their goals. Therefore, examining the results of a game-theoretic model tempers our expectations regarding our ability to achieve our national goals by reminding us that *the adversary always gets a vote.*[22] Simply because we intend for our actions to have a desired effect on an adversary does not make it so. Adversaries often have several ways to achieve their objectives, and blocking them from one will not prevent them from moving to their next best option. Only when we can shape the environment so that it is in neither side's best interest to attack will we achieve deterrence.[23] When deterrence fails, the game results show us the "best" outcomes that each participant can achieve given the capabilities at their and their adversary's disposal, where those capabilities are shaped by the investments made.

Investments in space capabilities take a relatively long time to pay off—multiyear investment horizons are common. Therefore, it was also important to the objectives of our research to use a methodology that would allow us to explore strategic interactions *over time*. To facilitate our understanding of how investments made today might shape conflicts in the future, the game we developed includes a multiyear period of investment prior to terrestrial conflict to gain insight into how nations might invest (or not) to position themselves for the coming conflict. Because we allow players to make decisions daily regarding whether to invest, attack, or defend, the decision to explore a ten-year period of interactions greatly expands the game space and computational burden but is essential for the research questions we pursued.[24]

---

[22] This quote is generally attributed to Sun Tzu in *The Art of War* and may be a paraphrase of his statement that "If we wish to wrest an advantage from the enemy, we must not fix our minds on that alone, but allow for the possibility of the enemy also doing some harm to us, and let this enter as a factor into our calculations" (Sun Tzu, *The Art of War*, trans. Samuel B. Griffith, New York: Oxford University Press, 1964).

[23] Similarly, when games result in wars of total destruction of the use of orbit, it is because it is either in (1) both side's best interest to destroy use of space for all to win the terrestrial competition or (2) one side's interest and the other is powerless to stop it.

[24] The game tree size is determined by the number of weapon/target pairs and attack intent, number of investments that could be made in each asset or weapon type, and duration of the game. Even when run on the most powerful parallel processors in the

As is the case in other analyses that use a game-theoretic approach, the adequacy and applicability of our analysis is limited by our ability to abstract the real context of the conflict, player motives, norms, rules, and myriad other operational factors into a collection of written rules and payoffs.[25] A successful abstraction allows us to discover nonobvious properties of the problem domain and develop an improved understanding of the possible conflict. In Appendix C of this report, we will discuss some of the more notable failures and successes we experienced in achieving the appropriate abstractions needed for our work.

Before proceeding further, it is important to note what game theory can and cannot do. We do not use our game to determine what an adversary *will* do. The objective of game theory is not to predict behavior, and, in fact, it is not actually capable of doing so—game theory solves for an optimal course of action, something humans rarely achieve. Furthermore, there are hundreds of parameters in our game, and it is unreasonable to believe that we can set all of them with enough accuracy to forecast a future competition. Game theory is a tool for normative analysis and is not predictive. The results of our analysis reveal what players *should* do, not *will* do, in response to specific changes in the overall competitive environment. In general, we found our game is best suited to the study of how asymmetries in the future battlefield might impact outcomes.

United States, reasonably complex symmetric games using our model take several days to solve (see Appendix B). Only in the last few years, with advancements in computing speed and parallel processing, has it been possible to apply game theory quantitatively to complex issues of multi-year conflict and cooperation. Joseph Y. Halpern ("Computer Science and Game Theory: A Brief Survey," in S. N. Durlauf and L. E. Blume, eds., *Palgrave Dictionary of Economics*, May 2007) provides a short tutorial on the challenges in solving complex game-theoretic models.

[25] See the discussion on implicit assumptions in game theory in J. D. Williams, *The Compleat Strategyst: Being a Primer on the Theory of Games and Strategy*, Santa Monica, Calif.: RAND Corporation, [1954] 2007.

Structure of This Report

In Chapter Two, we provide an overview of the game model, how we use it, and its limitations. We also provide generalize examples of "space war" principles we discovered from analysis techniques employed when analyzing game results. Chapter Three focusses on our research into the specific questions of the value of SDA-enhanced defensive capabilities and concealing or revealing offensive capabilities. Chapter Four describes the strategic interaction patterns that we found in the game. It also describes our attempts to discover the conditions that give rise to those patterns. Finally, Chapter Five collects our insights about the dynamics of space warfighting into a single summary.

Appendix A contains a chronological overview of our research since 2014 and places the content of this report within that larger context. Appendix B contains additional details of the game model beyond that provided in Chapter Two, with a concentration on implementation details of the parallelization of the game.[26] For those with an interest in the challenges of using complex models to perform research, Appendix C details some of our successes, as well as our failures, in developing the game structure, model, and analytic methods necessary to gain insight into our research questions. It concludes with a discussion of the lessons we learned about using AI to conduct research.

[26] Torrington et al., 2019.

CHAPTER TWO

Methodology

In this chapter, we provide an overview of the game model, some examples of how we use it, and its limitations. We also provide general examples of analyses of representative game play and the space war principles highlighted by those analyses.

Model Overview

Our model is a two-player game developed to help build insight into adversary behavior, strategy, and outcomes in a space conflict. We designed the model to allow us to explore a variety of potential research questions related to national investments that (1) improve the redundancy and resilience of space systems or (2) build offensive and defensive capabilities. The model plays out over an extensive time horizon (ten years) to examine how strategies manifest over time and in response to changing objectives. Thus, our model needs to include realistic representations of how adversaries perceive actions and the context of the conflict, how they reason, and how they make decisions. Capturing the richness and complexity of players is vital to the development of a realistic model of decisionmaking, and our model expands on traditional game-theoretic representations in four ways:

- Players may deviate from strictly rational valuation of outcomes based on their attitude toward risk (neutral, acceptant, or averse), which we term their *mindset*. Because the dynamics of conflict we explore in this report are only marginally influenced by player mindset, we do not discuss these factors in this report.[1] Additional details are described in Appendix B.

- Players may have multiple and asymmetrically weighted objectives, which we delineate using the various instruments of national power.[2] Yet even then we cannot model all the things a nation might value, and we discuss this limitation in the next subsection of this chapter. As Branislav Slantchev notes, the interpretation of a game-theoretic model is limited by the objectives modeled for each player. A game-theoretic model simply tells you "that if you consider these incentives and considerations important, then you might be interested in knowing they can interact in particular ways, resulting in some specific behaviors."[3]
- Players may have imperfect information about adversary capabilities. Chapter Three contains additional details about modeling this imperfect information and our modeling of decisions to reveal or conceal capabilities.
- Players make decisions based on a finite look-ahead. Rather than the perfect look-ahead common to many game-theoretic models, our players are limited to only being able to see six moves into the future.[4]

Figure 2.1 provides an overview of the basic elements of the game we just described.

---

[1] However, Triezenberg, 2017, finds that mindset is one of the primary determinants of achieving deterrence of the horizontal escalation of war into the space domain. All games reported here were played as both risk-neutral and risk-adverse/acceptant under prospect theory to assess whether mindset has a significant impact on our results. It does not.

[2] Mathematical representations of those objectives are provided in Appendix B. An in-depth discussion regarding why we chose these objectives is provided in Triezenberg, 2017. A discussion of how specific potential adversaries in space might weight these objectives in provided in Torrington et al., 2019.

[3] Branislav L. Slantchev, "On the Proper Use of Game-Theoretic Models in Conflict Studies," *Peace Economics, Peace Science and Public Policy*, Vol. 23, No. 4, December 14, 2017. For another perspective, see Ariel Rubinstein, "Comments on the Interpretation of Game Theory," *Econometrica*, Vol. 59, No. 4, July 1991.

[4] Moves may be fairly sparse, especially in games that result in deterrence, and often six moves cover the entire ten-year duration of a game. This limitation becomes more of a factor, however, in games of higher conflict intensity—i.e., in the games reported here.

FIGURE 2.1
Basic Components of the Game-Theoretic Model

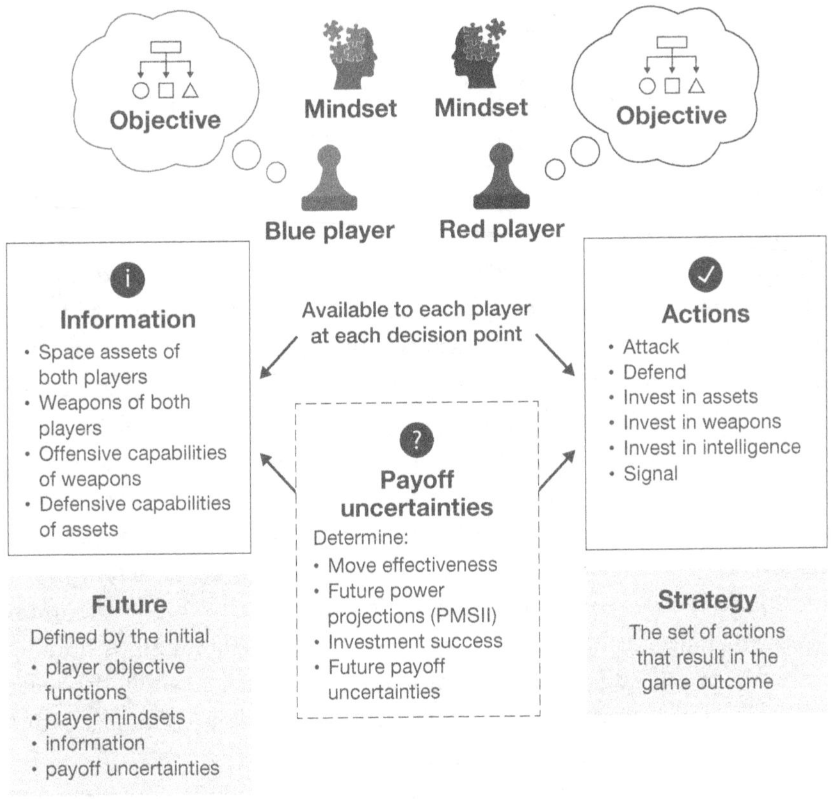

SOURCE: Adapted from Triezenberg, 2017, p. 69.
NOTE: PMSII = political, military, social information, and infrastructure.

Limitations of Our Approach

All Games Reported Here Are Between Near-Peer Competitors

Our game-theoretic model produces actionable insight into the dynamics of future space warfare only when we look at competition between near-peer competitors. Introducing any large asymmetry in the game results in either a game of total deterrence or total destruction. It is only when we can construct a base case that results in limited war between near peers that we can then vary the asymmetry of a particular attribute of interest and gain insight as to how that asymmetry affects game outcomes and possible conflict dynamics.

Invulnerability (or the Perception of) Creates Unstable Equilibrium

As just noted, our most productive investigations are when we construct the competition as being one between near-perfect peers and then vary just one other parameter to give one side an asymmetric advantage or disadvantage. This has an unfortunate side effect in that it can create an unstable equilibrium if the symmetry makes each side invulnerable. This is because the overall approach to "winning" the game is to find and exploit your opponent's vulnerabilities while shoring up your own. Games in which neither side begins with a vulnerability generally result in one of two outcomes—deterrence because neither side can effectively create an asymmetric advantage *or* wars of total destruction because each side perceives a means to achieve asymmetric advantage that does not in fact exist. In fact, however, there is great diversity in space capabilities, and perfect symmetry is unrealistic, making this unstable equilibrium, with its resultant all-or-nothing dynamic, possibly a false attribute of space warfighting. To study less-polar outcomes, we learned that we need to purposely start our games such that each player has a small vulnerability.

In modeling the United States, we chose to introduce the vulnerability of the Tracking and Data Relay Satellite System (TDRSS). Originally developed to provide near-continuous communication with space shuttle astronauts in low-earth orbit, TDRSS's satellites are placed in geosynchronous orbit where they appear to be stationary over a particular region of the earth. From this position, they can acquire and track satellites in lower orbits or even airborne vehicles, establishing a communication channel to relay information up from these overhead assets and down into U.S. ground stations using crosslinks. With recent growth in overhead imagery and reconnaissance data collection, demand for TDRSS services exceeds supply, meaning that any degradation in service may have an immediate impact on overall U.S. ability to observe and report what is happening in some contested part of the world. Inevitably, in our game runs, the adversary will try to exploit this vulnerability while the United States will invest to shore it up, starting a series of actions and reactions that then allow us to investigate how asymmetries in the game comparatively affect the game dynamics.[5] Without these initial vulnerabilities to start stable competition, it is much more difficult to determine the impact of the introduced asymmetries that are the subject of our research.[6]

Limited National Objectives

Our game is also limited in the number of items we used to characterize the objectives of the players and the overall state of conflict. The objectives are formulated as a weighting of a player's ability to project power (P [political], M [military], and SII [social, information, and infrastructure]), the value they place on possessing anti-satellite weapons, and the value they place on creating vulnerabilities in their opponent's ability to project M and SII power from space during a ground war.[7]

---

[5] As in our games, the United States is currently investing in ways to ensure robust access to overhead data-collection systems. Therefore, we might expect our adversaries to invest in ways to compromise that access or to find other ways to counter overhead collection.

[6] We create a similar small military vulnerability in each opponent's initial starting positions so that the overall symmetry of the game is preserved. Initializing the game in this way allows us to have greater confidence that the asymmetry of interest to the research, such as SDA-enhanced defenses or asymmetric possession of offensive weapons, is the dominate factor at play when analyzing game outcomes.

[7] Almost immediately, we realized that valuing creating vulnerabilities in an opponent's *political* capital led inevitably to wars of total destruction. Although fringe ideologies value wars of destruction to spur the end of time or the total overthrow of the world order, we were interested in modeling conflict and cooperation between near-peer nation states. To deprive an opponent of all ability to project political power would also deprive them of the ability to sue for peace and to ensure that terms of surrender are honored. Therefore, all games reported here had each player valuing their own political capital and showing indifference to their opponent's political standing in terms of national objectives. This is not to say their opponent's political capital did not figure into a player's decisionmaking. If each player knows that their opponent values their political standing, this valuation can then be taken into consideration when assessing how actions will be perceived by the opponent, driving the subsequent reaction. Triezenberg, 2017, discusses how powerful this valuation is, finding it to be a primary

Each player begins the game with an initial value for their political capital, and this political capital is maintained by avoiding the use of weapons that create debris or radiation. A player can accrue additional political capital if they are the victim of an attack on an illegitimate target.[8] All changes of political capital are fleeting and revert to the initial condition over time as outrage over the illegitimacy of attacks diminishes and what was once illegitimate becomes commonplace. This creates a use-it-or-lose-it dynamic in the game—players who have political capital take retaliatory action while world opinion is still in their favor, mimicking real-world political dynamics.

Each player also begins the game with an initial set of space assets that allow them to project M power from space or—in the case of nonmilitary satellite—SII. A player accrues power in these dimensions by building additional assets but only up to a threshold. Above that threshold, the value of additional assets is in the resilience they provide against attacks. Attacks on assets decrease their M and SII value, but, as with political power, these decreases are temporary—the model assumes that workarounds *will* be found, but there are specific investments the players can make to reduce the time it will take to recover their M and SII capabilities. We term *investments in resilience* as those that build capability above threshold or that help players recover capability faster; we find they are by far the most frequent investments made. The ability to project M and SII power from space is permanently reduced only when an orbit is destroyed by debris or radiation—after the *n*th attack of this type in an orbit, both players lose their ability

factor in achieving deterrence, an effect made more pronounced if national decision-making is influenced by prospect theory.

[8] Players can invest in diversifying their assets in a way that reduces the legitimacy of a target, thereby increasing their political capital if attacked. For example, a player can invest in hosting military communications on commercial satellites to create a mixed-use satellite that has less legitimacy as a target of war than a dedicated military communications satellite would. It is a subject of significant debate in U.S. space policy as to whether these types of mixed-use investments would be stabilizing or destabilizing. In our games, we did not detect a bias toward making investments of this type—perhaps the effects of political illegitimacy are outweighed by the attractiveness of the military target. Conducting specific research into the impact of these mixed-use satellites is recommended.

to project power from the orbit, setting up a mutually assured destruction dynamic within the larger strategic competition.[9]

Although possessing a varied arsenal of antisatellite weapons has its own intrinsic value in that weapons serve as deterrents and as a means of creating vulnerabilities in an opponent's M and SII, we also wanted to be able to simulate the dynamics of a space arms race. Therefore, our model includes an objective where players value the differential in how many different types of anti-satellite weapons they possess versus their opponent. This is the only zero-sum-game objective in our model of strategic competition.

Obviously, there are many other objectives that could drive national decisionmaking regarding cooperation and conflict in space, but we found these adequate to our research questions. From the beginning, we limited the weightings of player objectives to either 1 or 0—either the player cared about this aspect of national power, or they did not. Although more-nuanced weightings can be specified, we generally find the principles of strategic competition and cooperation are most observable when comparing and contrasting games where an objective is either fully present or removed entirely from a player's decisionmaking.[10]

Over time, we came to understand how each objective affects game play and were able to simplify further. Early on, we learned that, while valuing a differential in possession of anti-satellite weapons drove classic arms race

---

[9] We acknowledge that the debris threshold at which an orbit becomes unusable (or even whether such a threshold exists) is unknown. We included the debris threshold because we found it useful to our analysis to have a mechanism that mimics the dynamics of mutual assured destruction for space. The theory that there is a debris threshold at which cascading collisions makes an orbit unusable was originally proposed in Donald J. Kessler and Burton Cour-Palais, "Collision Frequency of Artificial Satellites: The Creation of a Debris Belt," *Journal of Geophysical Research*, Vol. 83, No. A6, June 1, 1978. This threshold for radiation attack is better characterized, unfortunately, with real-world data. In the early days of space (1962), the United States detonated a nuclear device in space (Starfish Prime), and, in 1963, up to one-third of the satellites in that orbit suffered catastrophic damage due to radiation effects on their electronics.

[10] The one exception was a group of games run for Triezenberg's dissertation in which she used a ratio of the players' dependencies on space to project power into a ground theater of war to weight the value each player put on creating vulnerabilities in their opponent's ability to project that power. Asymmetries in this balance were found to be a major factor in determining when deterrence can be achieved in game play.

behaviors in times of peace, it drove odd behaviors in times of war. Specifically, we noted that, if our players objectively valued a weapons differential in times of war, they would occasionally hoard weapons just to have them as opposed to hoarding them for later use. Although the second is a rational response to future uncertainty, the first was deemed irrational. Therefore, the players in the games reported here do not value weapons differentials in wartime—i.e., in times of war, the value of weapons is in their ability to hold adversary assets and weapons at risk, not in their mere possession. We were able to further simplify player objectives when we observed that differences in decisionmaking induced by separate M and SII objectives for power projection were small. Although a satellite's ability to project M versus SII (or a mix of the two) influenced decisions at the tactical level, none of our research findings was influenced by having separate valuations for M and SII. For our most recent investigations, we removed objectives related to SII power projection.[11]

All games in this report begin during peacetime, with players seeking to maximize their capacity to project military power from space, political standing in the world, and weapons capabilities relative to their opponent (i.e., winning the space arms race). The length of this period varies depending on whether the objective of our research is focused on the investments that lead up to the start of a ground war or on the dynamics post–ground war start, but, in all games reported here, it is set between five and seven years into the game. At ground war start, we change the objective—players no longer seek to maximize their relative weapons advantage and begin to value minimizing their opponent's ability to project power from space into the ground war. In other words, the space arms race ends, and the space power projection conflict begins. Deterrence is not an explicit objective of either player—it only arises if both players cannot improve their standing by engaging in attacks. At the same time, space war is also not an objective—players only attack if doing so will improve their overall score at the end of the game and will often attack before ground war starts if it is in their best long-term interest.

[11] Valuing SII power projection remains a capability of the game and may be of use in studying issues related to the increased entanglement of commercial and military uses of space.

Limited Parameter Sets

The parameter sets used for the game are extensive, yet we still acknowledge them as a limitation of the game. As our research progressed, we slowly expanded the parameters of the game to model aspects of reality needed to meet our research objectives. The complete parameter set for a game run has been split among seven files to simplify configuration management of game runs. The player names are generalized as Red and Blue,[12] and a weapons file and an assets file define the order of battle for each player. Each weapon is specified using 14 attributes, and each asset requires about 25 attributes to fully describe its capabilities.[13] A parameters file specifies probabilities of the kill chain,[14] legitimacy of weapons and targets, time to recover from attacks under various conditions, and investment timelines and payouts, among other details. Altogether, there are approximately 420 different values specified in the parameters file. All weapons, assets, and parameters just described are dynamic and change with game play as investments are made, attacks are executed, and defenses are deployed.

We also have files to configure more static aspects of the game. There is a file to define escalation thresholds and another to define the events that trigger changes in objective weighting for each player—i.e., redlines and deadlines. Yet another file defines the set of investments that each player is allowed to make. Investments in the game allow players to build or improve assets and weapons, collect intelligence that improves the various probabilities, and engage in deceptions, among other activities. Limiting the allowed investment types allows us to investigate such questions as, "What impact would it have if the United States has an asymmetric advantage in technol-

---

12 Traditionally in game theory, the Blue player is assumed to be the U.S. player, and the Red player is assumed to be the opponent. Because much of our work is theoretical and future oriented, we did not set parameters based on current U.S. capabilities, other than the TDRSS example given earlier to provide the necessary pretext to start the dynamics of competition. Similarly, we did not set Red capabilities to reflect any particular adversary. To achieve our goal of studying asymmetries, Red and Blue are modeled as peers in the space domain.

13 Assets possess varying types of defenses, hence the number of parameters needed to define an asset is variable.

14 *Kill chain probabilities* include probability of find, approach, effectiveness of weapon, and effectiveness of defense.

ogy X?" A final file defines the mindset for each player as a function of their risk perceptions for each national objective.

These files not only define the actual game truth for each player but also each player's perceptions of their opponent's objective value functions, mindset, assets, weapons, and capabilities. Structuring the initial conditions as different views of reality allows us to explore futures in which perceptions of adversary power differ greatly from actual power. See Figure B.3 in Appendix B for a pictorial representation of these different views of reality.

What we just described is only an introduction to the game to provide readers an understanding of its capabilities and limitations. More-thorough descriptions, including the mathematical equations that the game solves for, are documented in Triezenberg (2017) and Torrington et al. (2019) and are not repeated here. Significant changes to parallelize the game, undertaken since those reports were written, are documented in Appendix B.

Methods for Assessing Game Results

The research objective is to understand how investments and perceptions of investments might impact the strategies—i.e., the series of moves—used by players in the game. Therefore, for each solved game, the output is the time history of moves taken to invest, attack, or defend that optimize each player's objectives subject to the opposing player also optimizing their objectives. To provide additional visibility into the decisions that shape those time histories, we also capture the impact of those moves on each player's ability to project power from space and to satisfy their national objectives over time. Needless to say, in applying the game to any reasonably complex simulation of the United States and a potential adversary, the solved game produces results that are quite voluminous, and we need ways to synthesize and characterize the game outcomes if we are to find patterns that produce insights for policymakers. Although there are many ways to characterize strategy in a space conflict, we find the following metrics to be useful:

- Investment priorities: Although players choose between hundreds of different investments, to characterize the resultant strategy of each player, we binned the investments made based on whether they create or improve (1) redundancy and resilience of space assets, (2) offensive capabilities—i.e., the ability to hold the adversary's assets and weapons at risk, or (3) defensive capabilities.

- Attack timing: We look for correlations between the occurrence of first attacks and the overall attack strategy observed.[15] Specifically, we look for patterns in opening moves that might allow us to characterize strategies in space conflict similar to how strategies are characterized in games of chess.
- Attack legitimacy: Because the legitimacy of attacks heavily influences players' abilities to achieve their objectives and their opponent's sentiments in our model, we look for patterns in the types of attacks that are made. Two factors we specifically track are the number of debris-generating attacks and the number of reversible versus irreversible attacks.[16]
- Conflict intensity: We compute a conflict-intensity score based on both the number of attacks made and the length of time one or both players experience escalatory effects. Escalation increases based on each player's perceptions of attack legitimacy and the precedence of the attack and degrades over time if no further escalatory actions are taken.[17]

15 Overall attack strategies are discussed in Chapter Four.

16 An example of an illegitimate attack could be the degradation or destruction of a commercial asset (e.g., commercial communication or imaging systems). Reversible attacks generally use cyber- and directed-energy weapons to disrupt or degrade systems, but these weapons can also be used to destroy assets in a nonreversible way. More commonly, destructive attacks use radiation or kinetic weapons. It is the intent of the attack (e.g., disrupt, degrade, destroy) not the weapon used, that determines its reversibility.

17 Although the game allows players to have different perceptions of what is escalatory, all games reported in this report were run with shared perceptions of what is an illegitimate weapon or target. An interesting line of research would be to explore the dynamics of conflict when the parties have different perceptions of legitimacy and how that might give rise to unintended or accidental escalation. With recent events in the Persian Gulf and a rising interest in autonomous weapons, there has been speculation that unmanned systems, such as unmanned aerial vehicles or satellites, may introduce new and different mechanisms for unintended escalation. See, for example, CNA Center for Autonomy and AI, "Impact of Unmanned Systems to Escalation Dynamics," undated.

Having characterized game outcomes, we then use a variety of data analytics to (1) construct various visualizations of the variables in the game to gain insight into the factors that shape decisions, (2) seek out correlations between input parameters to the game and outcomes, and (3) perform comparative analyses between the time histories of similar games to identify whether there are moves that, when added to a player's possible set of moves, could be termed *game changers*. In the next few pages, we walk through examples of each of these analysis techniques to discuss what we can learn from them.

Comparative Visualizations of Game Results

A well-formulated comparative visualization of game results is perhaps the most direct way to shed light on how factors of interest to policymakers affect game outcomes. In his 2019 dissertation, Goirigolzarri displayed results using a simple but elegant visualization that scored game outcomes using a space-superiority metric.[18] This metric is defined as the integral of space operability over the period of the game, where Δt is the time step, T represents the period of the game, and *% functionality* is the space operability percentage at time t:

[18] Goirigolzarri, 2019. This dissertation also demonstrates the difficulty of interpreting a highly complex game. As acquisition timelines were varied, different optimal investment strategies emerged:

- If you are slow relative to your adversary, invest in simple resilience efforts and be reserved.
- If you are against a peer, invest in complex capabilities and be strategically assertive.
- If you are faster than your competitor, invest in complex capabilities and be reserved in conflict.

These strategies did not surface for every game played. Some games saw little discernible effect from shifting investment timelines. For Goirigolzarri's analysis, in fact, only ten examples in a 64-game experiment exhibited changes in game outcomes resulting from investment timeline shifts. The others were dominated by other factors that either led to games of deterrence or games of total destruction. In fact, over the course of our years playing the game, we found deterrence and total destruction were the two primary outcomes. Researching more-nuanced effects required careful construction of a base case that would allow us to study the smaller subset of games that were *not* all or nothing.

$$Space\ superiority = \int_{t=0}^{T} \%\ functionality \times t$$

To visualize and compare outcomes for both Red and Blue players, this space-superiority measure was plotted on a two-axis graph as shown in Figure 2.2, illustrating the space-superiority score both absolutely and relative to the other player for each game. Red's score is plotted on the *y* axis and Blue's is on the *x* axis. The diagonal line represents games where both lose equally. The shaded regions show who "won" the space conflict—i.e., who was better able to maintain power projection from space over the period of conflict and how dominant that victory was based on how far away the point is from the diagonal. In the plot, point A represents Red maintaining power projection virtually unscathed throughout the period of the game while Blue's power projection is significantly degraded. In effect, Red is winning the space conflict. Point B represents the inverse situation with Blue winning the conflict. Point C represents a game where both players lose significant power-projection capability, but Blue fares slightly better than Red.

This visualization can also tell us something about conflict intensity. In the upper right corner, both players' space superiority is maintained over the course of the game. These are largely games of deterrence—no player is winning, but, similarly, no player is losing. Movement down and left along the diagonal indicates that both players have lost significant space capability —i.e., game conflict was more intense. In this way, we can see that a game that resides at the top right corner would be a game of deterrence, whereas a game further down and to the left on the line would be a more escalatory game. In Figure 2.2, point E is more escalatory than point D. In fact, the best way to describe the game that resulted in point E is that both lose; this is the mutual-destruction scenario in which an entire orbit is destroyed by space debris.

Games played with realistic orders of battle against near peer adversaries of the United States fall very close to the diagonal line and tend to cluster either at the top end (deterrence) or at the bottom end (total destruction). **If space is to be a factor in the outcome of future wars against near peer**

FIGURE 2.2
Comparative Space-Superiority Scoring Visualization

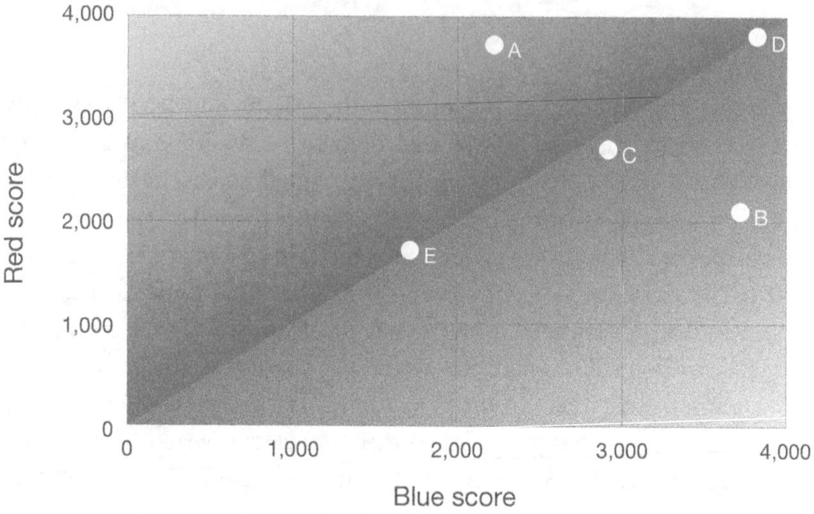

SOURCE: RAND generated for illustrative purposes only.

competitors, our work leads us to believe that it will be won at the margins. Space will matter only if one side or the other can better exploit temporary advantage.

Correlation Analysis

One of the simplest analysis techniques we use to examine the relationships between initial game conditions and game outcomes is linear regression; however, we learn very little from it. The game itself is not linear, and trying to understand it from a linear point of view yields only minimal insight. Appendix C provides an example of this technique as applied to the problem of misperceptions of offensive capabilities.

Given that the games we analyze hold all other inputs constant and we see low explanatory value for a regression correlating the one input we vary to our outputs of interest means that *other dependent variables must be emergent properties of the game*; i.e., if there is a linear relationship between a set of variables in the game and a measured outcome, such as conflict intensity, those variables are also outputs (not inputs) of the game. This highlights a key limitation in the use of linear regression for the analysis of a complex model, especially one that plays out over time—while perhaps at any given instant in time we could discover linear relationships between the conditions of the game and actions in the game, we only can catch a glimpse of relationships between starting conditions and overall game scores. Although regression analysis of our game input and output never produced a simple correlation (i.e., X leads to Y) from which one might formulate policy, it can and did often provide hints about the factors that matter in analyzing game results. For example, as we discuss in Appendix C, it alerted us to a fundamental difference between under- and overestimation of offensive capabilities.[19]

Game-Changer Analysis

Although correlation analyses can be used to examine the relationship between the capabilities that each player possesses at game start and game outcomes, it does not give us much insight into how those capabilities are leveraged to achieve the game outcomes. To gain insight into the latter, we examine outcomes as a function of the moves a player has available to them in the game. This is, by necessity, a complex analysis, but it yields insight and is invaluable as an aid to understanding how investments in both capabilities and in the TTPs to use those capabilities may play out in future space wars.

To understand why only some moves are available in the game, we first need to take a slight digression to understand the way our game is solved. In the games used in our research, there are roughly 200 moves available to each player, and players can play those moves on any given day of the ten-year period. A full game tree that attempts to solve all possible combinations of moves over time has trillions of branches and is computationally intractable.[20] However, we know that optimal solutions to the game contain only

[19] Prospect theory—a mindset that we apply to players in some of the games—does produce a bias about under-versus-over estimation. This is the bias that the regression analysis picks up.

[20] Even on the most powerful supercomputers available, such a game would take years to solve.

a handful of moves,[21] and we use this to our advantage to solve the games through a process of branch and bound, described more fully in Appendix B. For now, we will describe only the fortunate side effect of this branch-and-bound process. It mimics the way nations build capacity for war and gives us insights we never would have learned had the game simply solved for the optimal move set.[22]

In our game, each player begins with a set of weapons and assets that give them the ability to execute roughly 200 moves. However, all of these moves are not available at the start of the game; instead, each player begins a game with a set of three to seven moves that they have a priori determined to be their "best" moves without consideration of how their opponent might counter those moves.[23] This is not dissimilar to how nations build capacity—historians point out that outcomes in war can often be explained by the fact that nations invest in developing TTPs to defend or attack without adequately considering how an opponent could counter those moves.[24] The game is then solved with both players maximizing their objectives while subject to the other player also maximizing their objective and the scores are saved. Each player then tests each of their possible remaining moves to determine whether adding it to the move set improves their score in a fully solved game. Players take turns adding their most promising moves to their move set, and we solve and score the games after each selection of new "best" moves is made. When no player can add a move that improves

[21] Mathematicians would describe our game as having "few pure strategies" that optimize the game objectives.

[22] We are not the first to note that explainability is a necessary attribute of any AI tool. Had the game simply solved for the optimal moves and recorded them, we would have developed very little insight into the reasoning of why those were the optimal moves. Recall that our goal is not to find optimal investments—it is to understand how investments shape strategic outcomes.

[23] In addition to investing in new assets or weapons, other moves available to a player might be: "Improve the maneuverability of <asset>", "Degrade <asset> with <weapon> ", "Improve <asset> recovery after cyberattack", "Build redundant <asset> or <weapon>", etc.

[24] The French building the Maginot Line is a commonly cited example.

their score, we term the move set *optimal*.[25] We liken this process of adding moves to the move set to the real-world process of developing TTPs. It is not enough to invest in a new capability, players also need to invest in the TTPs to ensure that capabilities are used effectively.

A *game set* refers to the complete set of solved games over multiple iterations (each with the same initial conditions) as each player in turn adds moves to their move set. Often there is little difference in overall strategy and moves played over a game set—i.e., additional moves added after the first round of optimization yield only incremental change to outcomes. However, periodically, we observe a significant shift in game outcomes between two iterations of the same game. We term the moves added to the move set that result in these significant shifts to be *game changers*.

A concrete example may help explain the insights that can be developed by analyzing the shifts in game outcomes as moves are added to move sets and serve as a tutorial in analyzing the mutual optimization of game-theoretic models. Figure 2.3 shows the game-changer analysis for a game set in which Blue has a dominant order of battle. Two artifacts of the analysis are shown—the first is a plot of normalized final scores for Red versus Blue, the second is the list of moves available to each player during each game, with an X indicating the moves that are played. Each point on the scoring plot shows the outcome of a solved game, and the numbering indicates the sequence of that game within the game set. The same number is used to label the move set used in that game.

The game-by-game summary of the game set is as follows:[26]

---

[25] There still may be combinations of moves that would improve the game scores, so this move set is not, strictly speaking, optimal. It is, however, as close to optimal as we can get within the computational limits of today's technology and a reasonable time limit. As more moves are added to the games, they become more and more difficult to solve, and a game with set of 15 to 20 moves may take weeks to obtain results. We have come, however, to believe that optimality in the move set is not necessary for our work. Our research questions center on how investments shape the range of behaviors that could be expected in future space wars, and, for this reason, we view all games after each player has had a chance to add moves to their move set (i.e., above game 3) as valid sources of data on which to base observations.

[26] This game set is generated from an unclassified and purely fictional order of battle that we use as a regression test of updated versions of the game. Although designed with the goal of exercising a large subset of the functionality of the model—as opposed to

FIGURE 2.3

Example Game-Changer Analysis

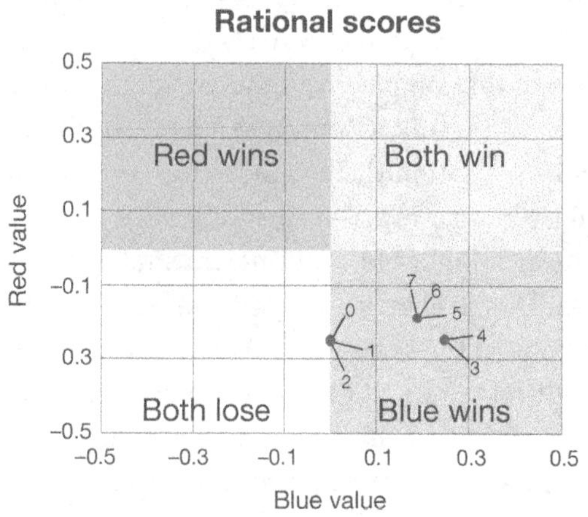

Move set								
1	2	3	4	5	6	7	Player	Moves
×	×	×			×		Blue	Discover opponent weapon capability red direct ascent ASAT.
×	×	×					Blue	Reveal weapon quantities blue cyber through supply chain.
		×	×	×	×	×	Blue	Blue missile degrade red ground jammer.
			×	×	×	×	Blue	Reveal weapon quantities blue ground jammer.
							Blue	Blue cyber through supply chain degrade red ground jammer.
						×	Blue	Blue cyber through supply chain deceive red ground jammer.
							Red	Fractionate service red relay satellite.
×	×	×	×				Red	Integrate use of allied assets red commercial SATCOM.
×	×	×	×				Red	Build weapons red cyber through supply chain.
							Red	Red ground jammer destroy blue commercial SATCOM.
							Red	Red ground jammer degrade blue military SATCOM.
							Red	Red cyber through supply chain destroy blue direct ascent ASAT.
		×	×	×	×	×	Red	**Build first-of-a-kind weapon red ground jammer.**
				×	×	×	Red	**Build weapons red ground jammer.**
						×	Red	Deceive asset quantities red relay satellite.

NOTES: (1) Each point on the scoring plot shows the outcome of a solved game, and the numbering indicates the sequence of that game within the game set. The same number is used to label the move set used in that game. (2) The zero/zero point on the scoring plot is the nominal score each would achieve by maintaining the status quo (i.e., if no moves were played). The axes are normalized such that −1 represents losing 100 percent of the value that could be achieved by maintaining the status quo, and +1 represents a gain of that same magnitude. (3) Regions in the scoring plot are shaded to denote outcomes where Blue, Red, or both "win"—i.e., they improve their score versus the status quo. Both can win because our games are not zero sum. (4) If a move is shaded red or blue in the list of moves, it means that the move was available to the player for use in the game. An X indicates that the move was played in the game. (5) Highlighted text in the list of moves indicates that it may be a game changer. ASAT = anti-satellite; SATCOM = satellite communications.

- **Game 1:** Examining the initial move sets, we see that Blue choses only three moves for their initial set, and that two of those moves—those to invest in intelligence gathering about the capabilities of Red's direct-ascent anti-satellite weapon and to reveal to Red the quantity of cyber weapons Blue has—are played. The third available move, the ability to kinetically target and degrade Red's ground jammer, is not played but is held as a deterrent. Recall that Blue has the dominant order of battle. This is reflected by their decision to select a minimal move set focused on (1) intelligence gathering, (2) removing any ambiguity Red might have regarding Blue's position of strength in cyber, and (3) deterrence.

  Red's initially selected move set is more extensive. They know they are disadvantaged by the starting order of battle and so chose a diverse set of moves that include investments in asset resilience and the building of additional weapons, as well as the ability to attack Blue's assets and weapons with their existing weapons. However, because Blue has successfully deterred Red from attacks, the only moves Red plays are the investments in resilience and building of cyber weapons.

  The score of this game is shown in the normalized plot—Blue successfully prevents Red from achieving about 28 percent of what Red could have achieved had the status quo been preserved while fully preserving Blue's own status quo valuation.
- **Game 2:** Blue is given the chance to add to their move set in the second game but chooses not to. Unless something else changes, no additional move improves Blue's position. Therefore, the move set and scoring of the second game is identical to that of the first.
- **Game 3**: This is Red's first chance to add moves to their move set. The move they add—building a new first-of-a-kind ground jammer—induces Blue to attack Red's existing jammers.[27] As can be seen in the plot, Red's score is not significantly changed, but Blue's score increases because, per their objectives, they are "winning" the arms race with

shedding light on a particular research question—it nonetheless provides a fascinating up-close look at how future space competitions could play out.

[27] In these games, the move sets themselves are not hidden. Each player knows what moves are available to their opponent. Only quantities of assets, weapons, remaining fuel, and other similar quantities can be hidden.

this attack on Red's weapons. Hence, we might term Red's added move a game changer. By building weapons, they have upset the status quo enough that Blue sees advantage in attacking those weapons.

- **Game 4**: In this subsequent game, Blue adds and plays a move to reveal the quantity of jammers they possess, but this does not significantly change outcomes. As modeled in our game, a reveal move has a 50-50 chance of being believed and even if believed, Red's new belief may have little impact on the game outcomes.
- **Game 5:** Red again has the opportunity to add moves and elects to add and play a move to build additional ground jammers. These additional jammers are game changers in that they give Red enough weapons capacity to survive Blue kinetic attacks (and in later games, cyberattacks) on those jammers.
- **Games 6 and 7:** Both Blue and Red add moves, but those moves are either not played or are deception/reveal moves that have no impact on the game dynamics. At this point, we terminate the game set. Clearly, the game set is reaching the point of optimality—neither player has a move that they can add that will make either player better off.

Although game-changing moves are *played moves* in these examples, this is not always the case. Examination over the range of games and their move sets reveals that game-changing moves often are *not* played. Instead, we find that merely the possession of the move changes the decision calculus for using other moves—it is the *potential* for these moves to be played that changes the game. This potential can act as a deterrent to the opponent or a reassurant[28] to the possessor; i.e., sometimes what changes the game is that possession of a move allows a more capable player to bide their time and not escalate a situation they might have otherwise felt compelled to contest.

A second caveat regarding game changers is that their importance may be overstated. In one of our investigations, we found an investment that provided Red a clear game-changing advantage over a range of initial conditions. Eager to understand this effect better, we ran that same range of

[28] This is a belief or action that deters aggression by reassurance; i.e. by convincing an actor there is no need to act because they will be able to meet their goals through other means.

initial conditions but with one key difference: We denied Red the ability to invest in the game-changing capability. Interestingly, although there were differences in individual outcomes, there were no statistical differences in outcomes over the range of initial conditions. Red simply found other next-best methods to use instead of the game-changing move we had observed in our initial runs. These next-best methods varied over the parameter space, emphasizing that **it is the plentitude of tactics available in the space domain that limits our ability to impact an opponent's strategic goals.** Although players prefer tactics that are effective over wider ranges of conditions, if a widely effective tactic is denied, they will find other ways to accomplish their goals. We found no silver bullets.

CHAPTER THREE

Strategic Value of Defensive Investments and Considerations for Revealing Offensive Investments

This chapter focuses on two specific investigations we conducted into questions about the value of (1) enhancing defenses and (2) concealing and/or revealing offensive capabilities.

Exploring the Value of Investments in SDA-Enhanced Defenses

To understand the value of defenses, we asked these specific research questions:

Assuming SDA improves a player's ability to defend against kinetic attacks, can those improved defenses deter the adversary from attacking?[1] *Do they change conflict outcomes?*

---

[1] See Triezenberg, 2017, for the applicability of theories of deterrence by denial of gains versus deterrence by threat of punishment to the space domain and for a historical overview of these theories of deterrence. Overall, she found investments in redundancy and resilience of space assets to be most effective in achieving deterrence, assuming that the player has at least some weapons with which to threaten a counterstrike. The question reported on here is much more specific: whether investment in a specific defensive tactic is likely to yield deterrence.

The United States is currently investing heavily in means to achieve better visibility into the physical objects that reside in space.[2] These objects include not just operational satellites, but defunct satellites, spent rocket stages, and debris of all sizes. Although a primary goal of this enhanced space situational awareness is to give operational satellites advanced warning of possible collisions with other objects, a secondary goal is to create a system for space *domain* awareness that can provide advanced warning against kinetic attacks on U.S. or allied satellites.[3] In light of this sizable and ongoing investment, we sought to understand the effect it might have on future space conflicts. Knowing that our game-theoretic model is best suited to the exploration of asymmetries, we gave our Blue player the option to invest in SDA while denying that same investment to our Red player. We hypothesized that we would gain insight into the value of SDA by observing

[2] These investments include improved ground radars and a series of satellites whose primary mission is to surveil the space environment. In low-earth orbit, the United States has deployed large telescopes looking up into the further reaches of space, and, at geosynchronous orbit, the United States has deployed a highly maneuverable set of satellites to patrol the orbit. The United States is also investing in "watcher" satellites that look down to detect incoming direct-ascent weapons. In addition, the private sector and U.S. allies have a number of sensors that provide additional data on objects in space. Incorporating all of these sensors into a comprehensive picture of SDA is an ongoing task for the USSF. For examples of these programs, see: Air Force Space Command (Archived), "Space Based Space Surveillance," webpage, March 22, 2017b; Air Force Space Command (Archived), "Geosynchronous Space Situational Awareness Program," webpage, March 22, 2017a; Theresa Hitchens, "Exclusive: NRO, SPACECOM Craft CONOPS for War in Space," *Breaking Defense*, May 4, 2020; Los Angeles Air Force Base, "Section 31: Bringing the Space and Missile Systems Center's Software Factory to Life," webpage, September 3, 2019.

[3] The term *space domain awareness* was adopted by the U.S. space community in 2019 to emphasize that, if the United States is to defend its assets in space, it needs to know much more than simply the position of objects in space and summary data about the purpose of those objects. To defend effectively, the United States also needs sufficient information to characterize threats in the space domain. See Sandra Erwin, "Air Force: SSA Is No More, It's 'Space Domain Awareness'," *SpaceNews*, November 14, 2019. The effectiveness of current U.S. investments in SDA against non-kinetic/non–ground-based attack vectors, however, is limited. Although domain awareness might give some early warning that a ground- or space-based jammer or laser/dazzler is in position to execute an attack, the attack itself occurs so quickly that reactions would need to be either preemptive or highly automated. Domain awareness of pending cyberattacks is very difficult to achieve and would require a wholly different set of investments.

(1) conditions under which Blue invests in SDA and (2) the conditions where Blue's possession of SDA has a deterrent effect on Red. To explore the latter, the base case for the research was structured around an order of battle that gave Red varying degrees of offensive advantage.

Research Approach and Game Design

For this research, we wanted to explore the defensive advantage of SDA—specifically, its ability to provide warning of a kinetic attack, giving the target an opportunity to mitigate or avoid the impact.[4] Therefore, we created a new object in the game whose only value is in providing SDA. That is, it is neither an asset that provides power projection into the ground war nor is it a weapon to be used in the space war.[5] By modeling these objects as purely providing SDA, any decision to invest in them and any attack on them is motivated by the advantage in defense they provide apart from any other objective the players might have.

We also modeled the effectivity of specific defensive actions, such as maneuvering or deploying near-field decoys to confuse an incoming kill vehicle—as a function of SDA. If SDA is below a threshold, the defense is less effective. If SDA is above that threshold, the defense is more effective. In other words, to reap the benefits of improved SDA, a player needs to *both* invest in SDA and subsequently take the defensive action. This allows us to segregate cases in which SDA is of no value because there are other effective deterrents and thus no need to take defensive actions versus those where deterrence has broken down but is restored by the possession of SDA. This model also allows us to compare the value of enhanced defenses against the cost of those defenses. This latter aspect is important because there is a penalty for taking active defensive action in space. For example, maneuvering a

[4] Another impact of improved SDA would be to allow a nation to better target their adversary's space assets and/or weapons. We did not study this aspect of SDA—we only addressed defensive improvements. Still another impact of improved SDA would be to provide more-convincing attribution of attacks in space. However, kinetic attacks are already relatively easy to attribute, so this effect was not studied.

[5] In real life, satellites can be used for multiple purposes. It may soon be usual for all satellites to carry a limited set of sensors that contribute to SDA. As noted elsewhere, almost any object in space can be used as a weapon.

satellite requires fuel and reduces overall satellite life. Furthermore, because most satellites are not designed to maintain service performance (e.g., high bandwidth communications or imaging) while maneuvering, defensive maneuvers may also reduce a nation's ability to project power into a distant theater of war for several days.[6]

Space Domain Awareness–Enhanced Defense Study Results

In examining game results,[7] we find that improvements in SDA-enhanced defenses fail to change game outcomes in any statistically significant way. *Our Blue player does not choose to invest in SDA even though it would result in an asymmetric advantage in defense against kinetic weapons.* This is true even when we have otherwise symmetric orders of battle—a condition in which we would normally find players searching out and exploiting even the smallest of asymmetric advantage. Furthermore, we observe that, *even when gifted with SDA at game start, our Blue player rarely takes defensive actions even though those actions would be highly effective against Red's kinetic attacks.* This is a puzzling result. It is only when we take a closer look at the weapons Red uses to achieve their outcomes that we notice the impact of Blue's investment in SDA. If Blue has SDA-enhanced defenses against kinetic attack, Red uses kinetic weapons less often. However, Red can substitute other weapons (in this case, an on-orbit jammer) to achieve the same strategic outcomes. ***Although the asymmetric possession of SDA fails to change Red's strategy, it does change the tactics by which that strategy is achieved.***

Figure 3.1 illustrates this result. Without SDA-enhanced defenses against kinetic attacks, Red quickly degrades Blue's military power projection to

[6] Similarly, decoys can usually be deployed only once, making them another nonrenewable resource. Furthermore, for decoys to be effective, it is likely that the satellite will need to match its electronic and thermal signature to that of the much simpler decoy, again taking the satellite out of service for a period. Gosnold, "Contested Space II: Countermeasures," SatelliteObservation.net, March 8, 2018, hypothesizes other passive and active techniques that could be used to defend against anti-satellite weapons.

[7] Note that only unclassified summaries of game results are provided in this report. Actual results have been archived on U.S. intelligence community servers.

FIGURE 3.1
Notional Illustration of Power-Projection Scores With and Without SDA-Enhanced Defenses

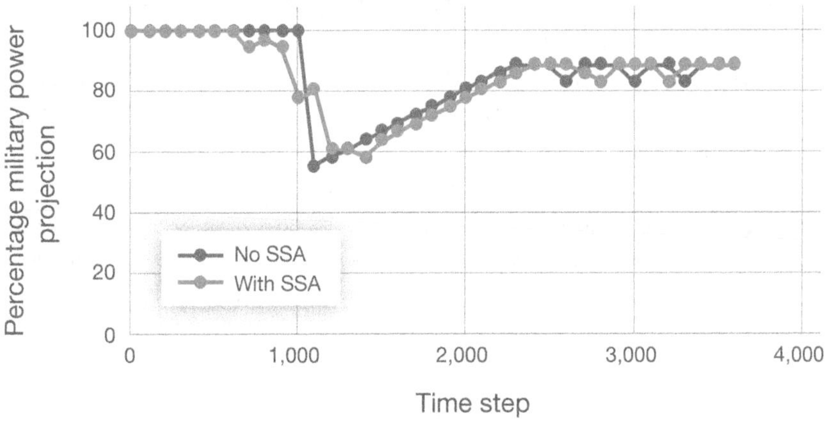

NOTES: This notional plot is illustrative of actual results from runs made at higher classification levels. The plot is notional but is based on actual dynamics observed in our research.

about 50 percent of capacity using a direct-ascent weapon. Blue then learns to work around that particular threat vector and recover military power, at which time Red switches to using cyber weapons to create smaller but still exploitable degradations in Blue's ability to project military power from space.[8] When Blue has SDA-enhanced defenses against kinetic attack, Red simply executes the same profile of attacks, but now the initial degradation of Blue's power projection is accomplished using a co-orbital jammer.

In fact, game outcomes for this series of runs show **no significant impact of SDA-enhanced defenses on U.S. ability to preserve military power projection from space**. Although adversary tactics are forced to shift, the overall strategic interaction pattern of the timing and severity of attacks is relatively unchanged. As a response to improved defenses against kinetic weapons, the adversary simply shifts its attacks to nonkinetic weapons.

8 As we will discuss later in the report, this is an "early shaping of the battlefield" strategy. The goal of Red's initial attacks appears to be an attempt to shift Blue's investments toward the goal of recovering military power-projection capabilities, which it successfully does shortly after the outbreak of the ground war.

Because there is no strategic impact, Blue is not motivated to invest in SDA and, if gifted with SDA, does not elect to use the enhanced defenses. There are other better uses for Blue's investment opportunities, and taking their satellites out of service to use the enhanced defenses would decrease their overall ability to meet their objectives.

It is important to note that although we did not find that SDA-enhanced defenses gave Blue better deterrence or better conflict outcome, it did shift the adversary toward less-destructive attacks. Kinetic attacks, with their propensity to generate vast debris fields, may be highly escalatory and threaten all nations' uses of space for peaceful purposes. *If investments in SDA-enhanced defenses serve no other purpose than to reduce the probability of debris-generating kinetic attacks on satellites, this is still a substantial benefit.*[9]

Lessons Learned

Importance of Accurate Modeling and Abstraction

A negative result, such as that found here, is very hard to understand. Only a model that cleanly segregates the targeting decisions of the attacker and the cost-benefit decision of the victim as they contemplate whether to engage in active defense against the attack can provide the insight needed to properly interpret the game results.[10]

To explore whether players who value the normative benefits of deterring debris-generating attacks might induce players to invest in SDA, we could develop an even higher fidelity model that includes reduction of debris as a separate objective.[11] However, despite improvements in run time gained by transitioning the game to a parallel supercomputing environment, any

[9] Although we did not give our players an objective of maintaining the use of space for future generations, reducing debris generation may be a policy goal that the United States should pursue.

[10] See Appendix C for a description of the process by which we arrived at the model used to generate the results described here.

[11] Our current model only incentivizes a reduction in debris-generating attacks indirectly via two effects: (1) the loss of political capital for the perpetrator of a debris-generating attack and (2) the cumulative destruction of the commons and the impact of that destruction on both player's ability to project power from space.

investigation we undertake requires a significant investment in researcher resources (monthslong) that we and our sponsor believe are better spent on other research questions. For now, it is enough to know that possession of SDA does result in fewer debris-generating attacks. Future research could be focused on determining if that outcome justifies the investments made.

Breadth of Tactics Available in Space Means There Are No Silver Bullets

This research reveals a clear principle of conflict dynamics in space or in any domain where there are multiple attack vectors: Improving defenses against one type of attack (in this case, kinetic) simply shifts the adversary's tactics, not their strategy. In response to improved defenses from additional SDA, the adversary duly shifted their attacks to nonkinetic weapons.[12] In other words, there is limited deterrence by defense in a conflict with a wide range of attack vectors. If there are multiple means to execute a strategy, the adversary thwarted from employing one attack vector will find alternate means. To change strategy through defense requires one of two things: (1) changing the objectives of the adversary or (2) identifying all of the ways the adversary's strategy could be executed and defending against each of those.[13]

Active Defense Can Be Counterproductive

A related principle illustrated by our exploration of this research question is that there is no deterrence by defense in a conflict in which deploying an offense comes with a relatively low cost to the attacker, while deploying a

[12] Later in this chapter, we describe results from games with imperfect information. Concealing the ability of SDA had no significant impact on game results or this finding. This is because although we can conceal the *source* of the enhanced defenses in our model, we cannot conceal the *effect* of the enhanced defenses. When a player finds a weapon to be ineffective, they will substitute the next best option.

[13] Game theoreticians assert that deterrence by defense is possible as long as the set of defenses is finite. This may be true in mathematical theory, but it ignores the reality that defenses are built through investments and nations have limited time and resources. Deterrence by defense is impractical in an environment in which any object can be a weapon. Spacecraft, although designed to meet the harsh thermal and natural radiation environment of space, are extremely fragile, and even a pebble can cause significant damage. See Reesman and Wilson, 2020.

defense require a high price from the defender. It is not enough to simply build systems that enable defensive actions—nations must also invest in ways to reduce the cost of employing the defense if they are to deter attacks. In the games we examined, the cost of taking a satellite out of service during the ground conflict—i.e., the certainty of incurring a self-inflicted wound—was rarely worth it when compared with the less-certain probability of kill. Furthermore, players rarely invested in improving those defenses in ways that would lower the cost. There were simply other better investments to be made.

Reflections on Reveal/Conceal of Hiding Defensive Capabilities

As just discussed, we wanted to understand how U.S. investments in SDA and in improved defensive capabilities against kinetic weapons might affect future games of cooperation and conflict in space. Unfortunately, the adversary responded to an investment in SDA by innovating in the development of nonkinetic weapons, and our investment or endowment in defensive capability resulted in a change in tactic but a relatively insignificant change in overall strategy and conflict outcome. Our work highlights that SDA on its own is insufficient as a deterrent. However, the possession of the potential to better defend against kinetic attack did shape adversary actions. It led the opponent to invest in and employ nonkinetic weapons. As we noted earlier, if this leads to fewer debris-generating attacks in space, the investment may be well worth making.

In contemplating these results, we note that one need not actually possess the ability to defend, it may be enough simply to convince an adversary that you have the potential to defend.[14] This highlights a key challenge in defining player behavior in these games. The adversary in the games reported above has perfect knowledge of how defenses are enhanced; they

[14] There is a wide body of literature on the deterrent value of cheap talk. We make no representations in this report about what might constitute a costly enough signal to convince the adversary that we have the potential to defend in space. However, we are pursuing this topic in future work.

know the improvement in defensive capability is only against kinetic weapons, and they know the cost that would be incurred if those defenses were employed. This adversary then explores all options to achieve their desired effect via alternative means. But this all-knowing adversary is unrealistic, and a more realistic representation of the adversary would include some perceived uncertainty about the effectiveness and cost of defenses. Uncertainty in effectiveness is hard to model in a game theoretic model. To do so would be to hide the impact of a move on the player's own objective. For computation reasons, we had elected not to fully model the fog of war; i.e., although we can hide many things in the game, both players always know the effectiveness of moves on their own objectives even if they do not know the source of that effectiveness or the effect of the moves on their adversary's objectives.

To illustrate this point, we conducted a series of runs to determine if hiding the SDA-enhanced defensive capability affects the game results. It does not. This is because the players can always observe outcomes—they know if the power projection into the ground war is or is not reduced by their attacks. So even if the source of the improved defense is hidden, they will still shift their attacks to nonkinetic means. Convincing the adversary that you have the potential to defend against a specific type of weapon may require more than words; i.e., a verbal bluff of defensive capability may not be enough, feinting may be required.[15]

The implication of the above discussion is that for the United States to effectively move our adversaries away from kinetic attack and into other, less debris-generating weapons, it may require a public demonstration of defensive capabilities against kinetic attack. However, we emphasize that this demonstration has a very specific objective: discouraging kinetic debris-generating attacks. Over all games we ran and all investigations we undertook, we find that there is a slight bias toward revealing defensive capabilities, but this is not a general rule. Our advocacy of demonstrations in this instance should not be construed as a recommendation that the U.S. space community should demonstrate all defensive space capabilities.

[15] Although *bluff* and *feint* are largely synonymous, a feint is a bluff that is conveyed through action. Paraphrasing from Merriam-Webster's definitions: To bluff is to present a false appearance; to feint is to deceive by a mock action.

Exploring the Strategic Value of Offensive Misperception

In this section, we discuss our investigation of these research questions:

Are misperceptions regarding an opponent's offensive capabilities (i.e., their ability to hold the opponent's space assets at risk) stabilizing or destabilizing? Do these misperceptions change conflict outcomes?

One of the impetuses for this investigation is that U.S. adversaries routinely assert that nontraditional U.S. satellites, such as the Geosynchronous Space Situational Awareness Program (GSSAP) or the X-37B space plane, are weapons.[16] U.S. writings on space also routinely point to nontraditional satellites from Russia or China as examples of possible weapons. In fact, almost anything in space *can* be a weapon, depending on how it is used. Therefore, our goal is to explore whether misperception of a nation's weapons capability is stabilizing; if this is the case, there is no need to clarify the purpose of these satellites. If the misperception is destabilizing, the United States might want to consider various ways to make the purpose of these satellites more transparent.

In our results between near-peer competitors, the effect of offensive misperceptions is quite nuanced. We found misperception to be stabilizing when (1) the adversary perceives a peer opponent to be disadvantaged or (2) a disadvantaged opponent is perceived to be a peer. In other conditions, we found misperception to be destabilizing or to have no clear trend. Decisions about whether to reveal or conceal investments in, or the actual extent of, offensive capabilities are not straightforward, especially when considering that not all potential opponents of the United States in space are peer com-

[16] GSSAP satellites support U.S. Space Command space surveillance operations and collect space situational awareness data to facilitate accurate tracking and characterization of man-made orbiting objects. See Air Force Space Command (Archived), 2017a. The X-37B orbital test vehicle, operated by the USSF, provides the ability to test new systems in space and return them to Earth, enabling more efficient development of space capabilities. See Space Force News, "Next X-37B Orbital Test Vehicle Scheduled to Launch," webpage, May 6, 2020.

petitors. Various levels of revealed information highlight the importance of strategic messaging in this round of games.

Research Approach and Game Design

To explore the effects of misperceptions of offensive capabilities, we conducted a series of game runs in which we endow a player with varying quantities of a GSSAP-like satellite and then vary whether that player considers them weapons and whether the adversary believes them to be weapons. We are interested in how adversary misperceptions of offensive capabilities—both the misperception that a player has weapons when they do not and the misperception that a player does not have weapons when in fact they do—impacts overall game play.

We observed in our prior work that deterrence breaks down when players have asymmetric offensive capabilities in games with perfect information. In other words, if either player perceives they hold an advantage, they take action to exploit that advantage. Parity, we found, is stabilizing.[17] Therefore, for this research question, we use an order of battle that favors Red, unless we endow Blue with a co-orbital weapon or, in the case of misperception, allow Red to believe Blue has co-orbital weapons. This endowment or perception makes the games more symmetric. Because of this, we might expect a more stable deterrence regime. There are theories in the literature, however, that both reinforce and temper that expectation. Scholars have written extensively about the ability of weapons to deter or escalate, pointing to the role of weapons and perceptions of these weapons in adversary decision-making. The works of Brodie and Schelling during the Cold War outline the stabilizing effect of weapons, while the work of Richardson presents a

[17] See Triezenberg, 2017. Parity as stabilizing is especially true when modeling decisionmaking using prospect theory. Prospect theory was developed in the field of behavioral psychology to address one of the unrealistic aspects of rational choice theory—the assumption that people make decisions based on a mathematical reasoning about the probabilities of success or failure in evaluating courses of action. Instead, prospect theory posits that people make decisions based on their perceived value as shaped by their risk tolerance. See Daniel Kahneman and Amos Tversky, "Prospect Theory: An Analysis of Decision Under Risk," *Econometrica*, Vol. 47, No. 2, March 1979, or Triezenberg, 2017, for a more detailed description of prospect theory.

logic model for the destabilizing effect of weapons build-up.[18] The deterrent effect of weapons is a central assumption of Cold War nuclear strategy. Brodie writes about the choice between seeking deterrence over assuring a winning capability, asserting that "[t]he objective of erecting a high degree of deterrence takes a higher priority than the objective of assuring ourselves of a winning capability, if for no other reason than that the first is likely to be a prerequisite to the second anyway, and is likely to cost less."[19] Schelling discusses the role of weapons not in their ability to destroy, but their ability to deter, asserting that "[t]o inflict suffering gains nothing and saves nothing directly; it can only make people behave to avoid it. The only purpose, unless sport or revenge, must be to influence somebody's behavior, to coerce his decision or choice."[20] Both Brodie and Schelling therefore illustrate a logic model in which the presence of weapons can be stabilizing by promoting deterrence.

However, other theories emphasize that increased weapons stockpiles are destabilizing. Lewis Richardson postulated that the likelihood of two nations engaging in a conflict increased with increasing weapons stockpiles (or the ability to project armed strength) and was tempered by domestic resistance to the conflict, illustrating the role of investments in weapons during an extended competition.[21] He developed a set of differential equations to determine the growth of these stockpiles. Each equation represents the rate of change of arms buildup, with this rate increasing with such factors as fear, ambition, and revenge and decreasing with such factors

---

[18] Bernard Brodie, *The Anatomy of Deterrence*, Santa Monica, Calif.: RAND Corporation, RM-2218, 1958; Thomas C. Schelling, "The Diplomacy of Violence," in John Garnett, ed., *Theories of Peace and Security*, London: Palgrave Macmillan, 1970; Lewis F. Richardson, *Statistics of Deadly Quarrels*, Quincy Wright and C. C. Lienau, eds., Pittsburgh, Pa.: Boxwood Press, 1960.

[19] Brodie, 1958, pp. 9–10.

[20] Schelling, 1970, p. 65.

[21] Lewis F. Richardson, Nicholas Rashevsky, and Ernesto Trucco, eds., *Arms and Insecurity: A Mathematical Study of the Causes and Origins of War*, Pittsburgh, Pa.: Boxwood Press, 1960.

as restraint or fatigue.[22] Figure 3.2 illustrates this set of Richardson's equations. An examination of the math shows that if x and y are equal (weapons parity) and the cost of defense is equal to the cost of offense ($a = m$ and $b = n$), then nations will not build arms unless their sense of grievance is high. Because we have no grievance in our games, we would expect that, with weapons parity, nations will invest in building weapons only if the cost of building a defense is more than the cost of building offense—and in fact we do see this effect in our games.[23] Although Richardson's equations can be used to predict the growth of the stockpiles, at other times in his writings, he describes x and y in the equations of Figure 3.2 as the propensity to use armed strength. Whether he meant for weapons stockpiles to be interpreted

FIGURE 3.2

Richardson's Equations Illustrate Conditions Leading to an Arms Race

$$\frac{dx}{dt} = ay - mx + r$$

$$\frac{dy}{dt} = bx - ny + s$$

- x and y represent weapon quantities for each player
- a and b are the "defense coefficients" of each player
- m and n are the "fatigue and expense" coefficients
- r and s are the "grievance" coefficients

22 Richardson, 1960, and David Bigelow, "An Analysis of the Richardson Arms Race Model," November 25, 2003.

23 This effect is obscured in our games by the dominating issue of the cost of using defensive techniques for satellites. Richardson's equations do not consider the cost of taking an asset out of service to execute a defense, but only the cost of developing the defense. For his 1950s analysis, this was sufficient. In fact, the cost of executing a defense—a compelling factor in space war—has few analogies in more conventional domains of warfare. An analogy from conventional warfighting might be a decision to keep a troop garrisoned temporarily to preserve its ability to fight at a later time. Few would call garrisoning a "defense," but many defenses proposed for use in space have this same effect.

as increasing the probability of their use or whether he simply used the same equations for two different phenomena is debatable.[24]

Richardson's theory of war as the outcome of a system of differential equations is at odds with the game-theory concept of war used by Brody and Schelling. Anatol Rapoport makes the case that both methodologies have a place in modeling war and that much of their limitations could be relieved with better computational power that could solve complex nonlinear systems of equations, which today's game-theoretic models allow.[25] Therefore, it is reasonable to expect that we might find evidence of weapons possession as either stabilizing or destabilizing in our games.

In summary, from both our prior experience using the game and from the literature, we expect the development of weapons, or the perceptions surrounding such development, to significantly affect strategy and outcomes in our game. With the goal of designing a set of games that might begin to untangle these theories, we designed a set of runs where a GSSAP-

---

[24] Richardson was a pioneer in using differential equations to describe the possible origins of war. Therefore, it should not surprise us if he used a generic system of equations to explore multiple phenomena. In *Statistics of Deadly Quarrels*, he uses the equations in passing to demonstrate the propensity for war; however, none of the statistics he provides support the idea that larger weapons stockpiles lead to more wars. Instead, he finds the frequency of war to be positively correlated with the number of states with which they have common frontiers, a desire for revenge, and economic causes and negatively correlated with the length of time the parties have participated in a common government. In fact, when Richardson performed correlation analysis, he did not find a statistical relationship between armed strength and outcomes of war, a relatively underappreciated aspect of his research (Richardson, 1960). A possible reason for the ambiguity about whether Richardson actually believed that weapons stockpiles increased the propensity for war is given by Anatol Rapoport who, when writing about Richardson's equations in 1957, notes that the theory that armed strength is the primary factor in deterring war was the "de facto" official view of the United States that "may not be opposed or even questioned in any context" (Anatol Rapoport, "Lewis F. Richardson's Mathematical Theory of War," *Journal of Conflict Resolution*, Vol. 1, No. 3, September 1957).

[25] According to Rapoport, 1957, p. 90,

> Possibly the true nature of large-scale human events is intermediate between physical determinism and cognizant choice based on the evaluations of potentialities. If so, then the findings of game theory and systems of differential equations are the two extremes bracketing the yet unknown theoretical method suitable for the study of human behavior.

like satellite may carry a weapon. These weapons can be endowed initially and/or the players can elect to invest in building the capability. Additionally, their investment in the weapon can be hidden and/or revealed, as can the quantity of the initial endowment. It should be noted that both players are also given other weapons, so this analysis is not whether one has any weapons at all,[26] but whether deception about the quantity and type of weapons changes game outcomes. In later runs, we deliberately chose a weapon capable of reversible attacks (i.e., an electronic warfare [EW] weapon) to ensure that differences seen in the games are not confounded by the stigma associated with kinetic debris generation or with nuclear radiation.

Offensive Misperception Study Results

On average, the game outcomes did not change based on whether the possession of weapons was hidden or revealed. Nor did they change based on the quantity of weapons that were hidden.[27] Furthermore, for reasons we discuss in Appendix C, players in a game-theoretic model do not play moves to "reveal" or "conceal" in a realistic manner. Although correlation analysis indicates that there might be a relationship between concealment and conflict intensity (see discussion of regression analysis in Appendix C), there is no overall correlation of outcomes to input conditions. Additionally, the perception of the quantity of weapons does not consistently lead to the same results as when the quantity is known. Deception makes a difference, but averages and correlations do not yield a consistent pattern. In short, to build a narrative and better understanding of game play, we need a hypothesis to test.

---

[26] The number of nations that have demonstrated the ability to deploy ground-based kinetic attacks against objects in low-earth orbit is steadily growing. To date, the United States, China, and India have all demonstrated the capability to use a ground-based missile to destroy their own satellites in low-earth orbit, and Russia is known to have the capability. Any nation with access to reasonably good technology can convert a conventional ballistic missile capability to be a direct-ascent weapon.

[27] When the quantity of weapons was revealed, the impact was as documented in Triezenberg, 2017: Under prospect theory, parity of weapons is stabilizing independent of quantity of weapons while under rational choice, quantity of weapons is destabilizing independent of parity.

Such a hypothesis surfaced while discussing our correlation observations and various theories of conflict with the sponsor during bimonthly interim reviews of the research. The suggestion was made that our observations might be a manifestation of Sun Tzu's admonition, "when weak appear strong and when strong appear weak."[28] Focusing our analysis on this hypothesis, we compare cases of deception with cases of complete information about the status of Blue's weapons stockpile and find a compelling case for the relationship between Red behavior and their perceptions about Blue weapons. More specifically, we compare the results from different scenarios to determine how Red perception shapes the game results:

- Blue's possession of co-orbital weapons, or lack thereof, is common knowledge (truth).
- Red overestimates Blue's possession of weapons (when weak, appear strong).

TABLE 3.1

Conflict Intensity as a Function of Deception Regarding Weapons Possession

Blue Possession of Co-orbital Weapons	Adversary Perception	Ability to Maintain Blue Power Projection from Space
Has 0 or 1 co-orbital, compared with Red's 6 co-orbitals	Overestimates U.S. offensive capability	Deterrence
Has 0 or 1	Knows truth	Limited war
Has 3 or 6 (i.e., parity or near parity with Red)	Knows truth	Limited wars with some wars of total destruction
Has 3 or 6	Underestimates U.S. offensive capability	Deterrence

NOTES: Colors represent the average conflict intensity score over all games of this type. Green = lowest intensity; yellow = medium intensity; red = highest intensity. This table includes cases of differing player mindset, and, in some games, there are additional differences in the order of battle that may favor Red (i.e., parity in co-orbital weapons does not necessarily imply overall parity).

[28] "Hence, when able to attack, we must seem unable; When using our forces, we must seem inactive; When we are near, we must make the enemy believe we are far away; When far away, we must make him believe we are near" (Sun Tzu, 1964).

- Red underestimates Blue's possession of weapons (when strong, appear weak).

Table 3.1 shows a comparison of conflict outcomes. It is clear that either underestimation or overestimation provides better deterrence and less-intense conflicts than when weapons possession is known. The comparison does not, however, provide us much insight about why.

To explore why this pattern may exist and to ensure that it is not influenced by the stigma associated with a kinetic weapon, we constructed unclassified games where our players are at parity, with the exception of Blue's possession of co-orbital EW weapons. It is not enough to examine any single game setup and simply vary whether the quantity of these weapons is known to Red. The pattern only makes sense when we look at all four variants of whether Blue does or does not have this offensive capability and whether it is revealed or concealed. To illustrate this point, we next examine in depth just two of the four games. In these two games, we do not give Blue any EW weapons at game start, but Red has such weapons. We then vary Red's perception of whether Blue does or does not have that offensive capability and look for differences in the games:

- Game scenario 1: Blue has no EW, and this is common knowledge (i.e., both know that Blue is at a significant weapons disadvantage).
- Game scenario 2: Blue has no EW, and Red thinks that Blue does (i.e., Red thinks the players are at parity, but Blue knows their own weapons disadvantage).

What we observe from these games is that when Blue starts with a weapons disadvantage, regardless of what Red believes, Blue decides to initially invest in EW weapons to rectify the weapons disparity. After the initial weapons build, Blue switches to investments that enhance the resilience of its vulnerable asset either by increasing redundancy or by decreasing the asset's fail-safe and recovery time.[29] Although this Blue investment strategy

[29] Recall from our discussion in Chapter Two that both sides start with one asset that has a small vulnerability. This is done to create a background of stable competition against which to judge the results of asymmetries between the players.

is relatively consistent between the two scenarios (Red knows Blue has no weapons versus Red does not know), the attack patterns of the two games are dramatically different. When Red perceives that Blue has weapons and the two players are at parity, Blue and Red both engage in relatively minor attacks, and deterrence is achieved. However, when Red knows that Blue does not have EW weapons and thus is at a disadvantage, both sides ultimately engage in a war of total destruction. This is exactly what we would expect to see in cases where parity or disadvantage are real, not merely perceived—players at parity will exchange tit-for-tat blows, but the deterrence regime is stable, and these attacks do not escalate. However, when a player perceives advantage, they will attack, and, if the disadvantage is large enough, the disadvantaged player may find that assured destruction of space for both is their best outcome. In short, deception of reality seems to have the same effect as reality.

It is only by examining all four cases—Blue is at a disadvantage and Red knows or does not know, and Blue is at parity and Red knows or does not know—that we can begin to understand that perceptions of reality do not always produce the same impact as reality. Panels A–D in Figures 3.3–3.6 illustrate the point. In this series of graphs, we have plotted Blue's ability to use their assets to project military power, *M*,[30] over the ten-year period of the game.[31] In all panels, the ground war starts at day 2000.

- Panel A is the case where Blue hides the fact of weapons disadvantage from Red (i.e., scenario 2, stated earlier). Red initiates a few attacks, but Blue plays tit-for-tat with the weapons they do have, and escalation does not occur.[32]

[30] This *M* is based on the quantity Blue's assets and the criticality of those assets to military power projection.

[31] The pattern would be equally clear if we had plotted Red's power-projection capability: When there is deterrence, neither side sees a degradation in power projection; when there is total war, both lose their ability to project military power through space.

[32] A tit-for-tat strategy in our game does not mean that both players use the same type of weapons or attack the same type of targets. It means that attacks are proportional in terms of their impact on each player's objectives. This may not be true in real life. We have observed that, despite it being policy that the United States will respond to attacks "in the time and with the means of its choosing," public perceptions of proportional-

FIGURE 3.3

Panel A: Red Perceives That Blue Has Parity, When Blue Does Not, Is Stabilizing ("When Weak, Appear Strong")

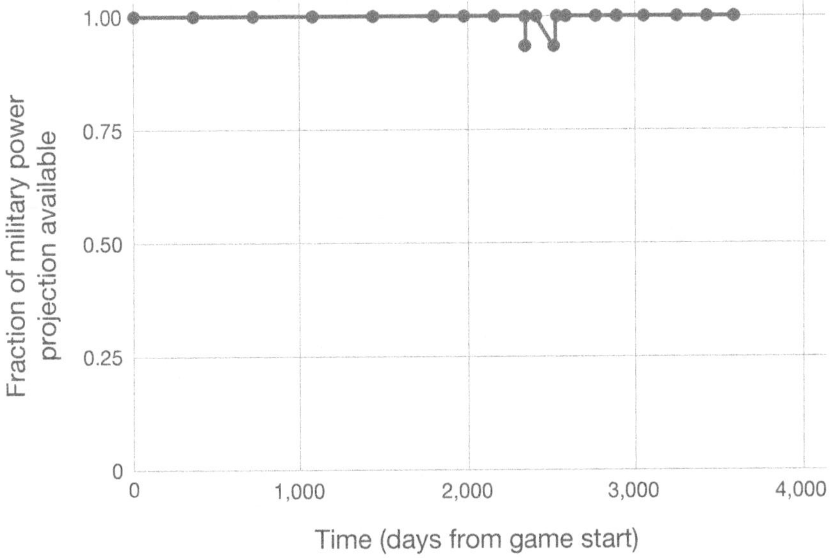

FIGURE 3.4

Panel B: Red Perceives That Blue Is Disadvantaged, When Blue Is Not, Is Stabilizing ("When Strong, Appear Weak")

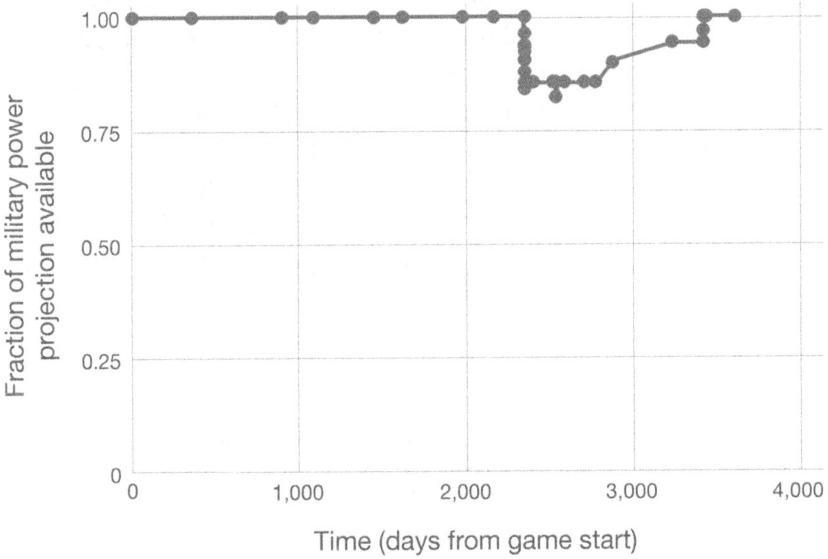

FIGURE 3.5

Panel C: Complete Information About Blue Weapons Disadvantage Is Destabilizing (Orbit Is Destroyed, Both Sides Lose Use of Space)

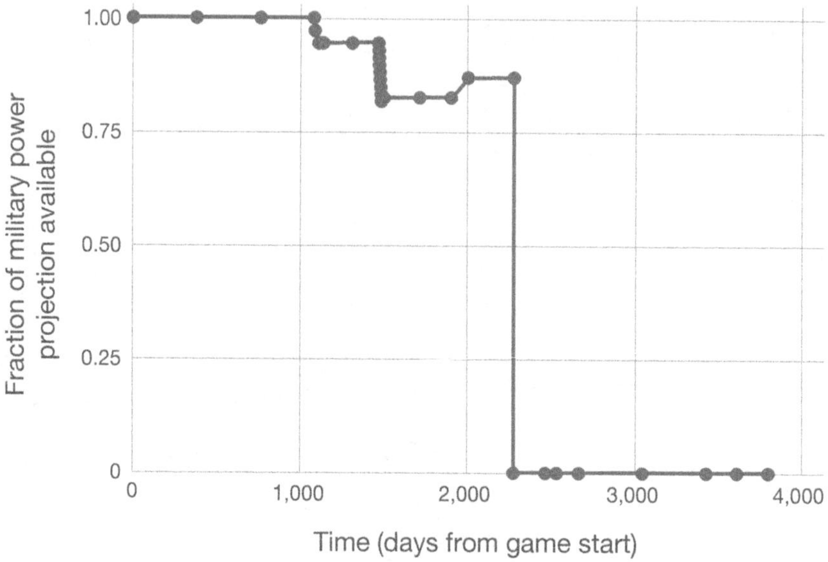

- Panel B is also a deception case, but, in this game, Blue has parity in weapons, but Red does not know it. In other words, Red has underestimated Blue's possession of offensive capability and believes in their own advantage. As Panel B shows, Red initially attacks quite strongly to press this perceived advantage, but quickly learns by Blue's response that attacks will be reciprocated. The net result in both cases, whether Red merely believes there is parity or whether Red becomes convinced of parity through Blue's response, is the same. Neither Red nor Blue find it in their best interest to continue attacks, and, after a series of tit-for-tat attack and response moves, the space war does not escalate.

---

ity are highly influenced by the similarity of weapon/target pairings. Responding to an attack on a spacecraft by bombing an opponent's military base may not be perceived as proportional even if neither result in loss of life.

FIGURE 3.6
Panel D: Complete Information About Parity in Weapons Is Destabilizing (Orbit Is Destroyed, Both Sides Lose Use of Space)

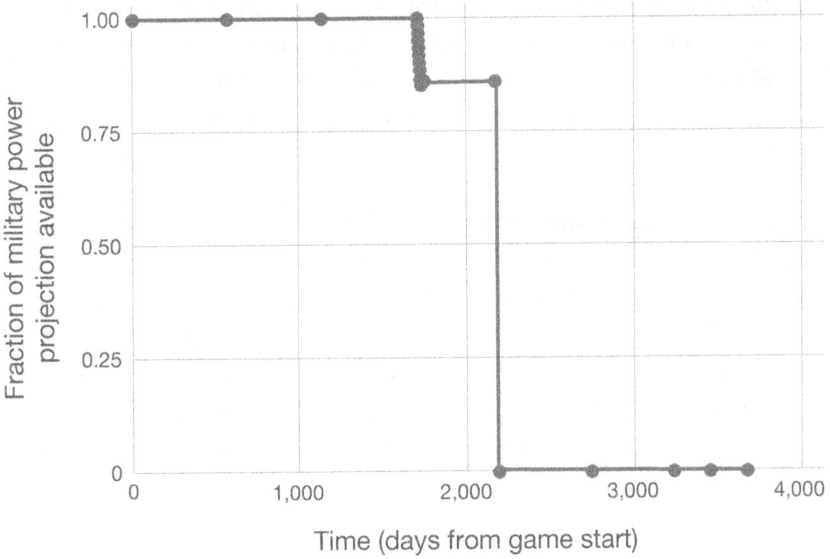

In both cases, deception has added enough uncertainty to the game that either the perception of parity in offensive capability or uncertainty about the source of actual parity is stabilizing. This is very similar to Triezenberg's finding that the unstable equilibriums that exist when players are at parity can be stabilized under prospect theory—even small forces that act against strictly rational decisionmaking are stabilizing.

Panels C and D tell a different story. Although both cases where truth is known lead to wars of total destruction, the dynamics are slightly different.

- Panel C is the case where Red has complete knowledge of Blue's weapons disadvantage (scenario 1, stated earlier). In this case, Blue's early build of weapons creates a use-it-or-lose-it scenario for Red. Red engages in early attacks to shape the battlefield (we will say more on this strategy in the next chapter) before losing weapons advantage. In fact, Red's attacks push the orbit nearly to the brink of destruction, and, after the ground war starts, Blue executes attacks that destroy the orbit, ensuring that both lose. This should not be surprising—without depth in their arsenal, a both-lose scenario is the best outcome that Blue can accomplish.

- Panel D is the case where Red and Blue have weapons parity, and both know it. Neither have advantage, but both are looking for even the smallest advantage and have plenty of weapons to waste. A war of attrition occurs almost simultaneously with the ground conflict and quickly escalates to total destruction. There is very little strategy in this game.

These plots imply that if Blue has offensive parity, hiding this capability from Red will have a stabilizing effect (where we define *stabilizing* as resulting in a lower conflict intensity). Similarly, if Blue does not have offensive parity, hiding this fact from Red will also be stabilizing. In short, the results of this analysis affirm Sun Tzu's assertion that hiding offensive capabilities, or lack thereof, results in maintaining power-projection capability and enhances the ability to deter the horizontal escalation of ground wars into the space domain.

Implications for U.S. Space Policy

The results of these experiments identify how a nation's ability to project military power from the space domain throughout the time frame of the game is affected by misperceptions about offensive capabilities. We can see from the results a clear advantage in military power projection when offensive capabilities are hidden, both when the adversary overestimates and underestimates. These results corroborate Sun Tzu's admonition: "when weak, appear strong; when strong, appear weak." This hide/reveal experiment highlights two important considerations in policymaking for the space domain. One is that the United States has the ability to shape adversary decisionmaking through its investments and possibly through what it chooses to reveal or conceal about those investments. The other is that the optimal message depends on the adversary, and the benefit of such messaging will rely on our understanding of the adversary and our ability to control this message for each potential adversary.

Regarding whether the United States should make greater efforts to clarify the intent of GSSAP and the X37-B, uncertainty appears to be generally stabilizing if our concern is a future conflict with a near-peer adversary. If U.S. concern is instead a conflict with a less-capable adversary, we believe from our prior work that other factors, such as dependency on space to project power into ground wars, will dominate in determining whether the United States can deter the escalation of that conflict into the space domain.[33] Revealing or concealing offensive capabilities may make no difference. Therefore, overall, we believe the United States should continue to operate these satellites in a nonthreatening manner to avoid creating a use-it-or-lose-it scenario for possible U.S. adversaries. Furthermore, we believe that it is not necessary, and may even be detrimental, to remove all ambiguity about the capabilities of these satellites. Overall, our research finds a slight bias toward revealing *defensive* capabilities, but we find a strong bias against revealing *offensive* capabilities.

---

[33] Triezenberg, 2017, varies dependency on space to illustrate that it may be impossible to deter a player who is less dependent on space from horizontally escalating war into the space domain.

CHAPTER FOUR

Characterizing Strategic Interaction Patterns

In this chapter, we discuss approaches to understanding the following research question:

When conflict in space occurs, do the players engage in distinct attack strategies similar to those observed in games of chess or other strategic competitions?

This phase of the research was motivated by our observation after playing more than 100 games over the course of several years. Games—other than those that resulted in deterrence or total war—seemed to be falling into distinct patterns.[1] In this chapter, we discuss characteristics of each of these patterns and some observations about them. We also include a short recap of our efforts to understand how U.S. investments and the revelation or concealment of those investments might determine which of these strategies is used in a future conflict that extends into the space domain.

1 For clarity, these strategies are in addition to those in which both sides maintain the status quo (deterrence) and those where both sides engage in a war of destruction. Although those games are interesting, prior publications have dealt with the conditions that we found give rise to them. In this report, we focus on conflict dynamics short of total war.

Finding Strategic Interaction Patterns

In game-theoretic models, players have strategic goals that are expressed mathematically by their objective function. Each player then selects actions that optimize their objective function, subject to knowing that the opposing player is doing the same. In the broadest interpretation of game-theoretic modeling, a *strategy* is defined to be the series of moves players take over the course of the game. Yet looking at series of moves did not help us to identify strategic interactions patterns that might be meaningful to future warfighters. As we discussed earlier, there are many ways to accomplish a strategic goal in space. The wide diversity of tactics available to the players means that a series of individual moves does not yield strategic insight. As discussed in Chapter Three when discussing the impact of SDA-enhanced defenses, only by looking at the impact of those moves on military power projection were we able to observe that depriving an opponent of a particular tactic had not changed the opponent's ability to carry out their overall strategy.[2] Furthermore, as discussed in Chapters Two and later in Appendix C, attempts to reason about patterns seen in the games by computing regressions against initial conditions did not lead us to an understanding of strategy. Instead, we found insight only when we had a hypothesis, or principle, to guide our analyses. Clearly, a wider view of strategy would need to be taken.

Given these observations and knowing that the set of strategies that can be played in chess and go are often constrained by the opening set of moves, we decided to look in detail at the opening moves in our games.[3] Unlike the game-changer analysis we described in Chapter Two that looked at available moves to players and what moves were played, we now looked at *when* moves were played. We took careful note of whether opening moves were invest-

---

2 Although our SDA analysis found that depriving an adversary of one tactic did not change their ultimate strategy, there was nothing in our analysis of the defensive advantages of SDA that led us to an understanding of *why* the adversary had chosen that particular strategy.

3 This is not akin to a principle in game theory called *first mover advantage*. In a game of repeated interactions, such as we have modeled here, we observed no clear advantage in being first to move. In fact, we find that the value of any particular move is highly dynamic. The advantage goes to the player who makes the right move at the right time, not to the player who is first to move.

ments in weapons versus resilience versus building or improving defensive capabilities and which player made them first. We also noted which player was the first aggressor and the timing of first significant attacks.[4] Although our examination of opening moves in and of itself did not yield results, it did bring to our attention a set of games in which players engaged in offensive attacks well before the ground war broke out, sometimes by as much as two or three years. This was curious to us because at the time these attacks are made, the player's objective functions do not value the creation of vulnerabilities in the opponent's power projection from space.[5] Basically, players were attacking when those attacks had little or no payout in terms of changing their scored objectives. Why, we asked, would a player sacrifice weapons in this manner for no immediate payout? Clearly, it must be because these attacks allow them to achieve a better score over the course of the future conflict—i.e., these players must be playing a long-term strategy. We came to call games that exhibit this dynamic "early shaping of the battlefield." We can construct a reasonably well-validated theory of warfare regarding this strategy. In many of these games, attacks appear to be an attempt to shift the opponent away from investing in weapons and compel them to instead invest in recovering the space capabilities lost in the attack. By attacking early, a player forces their opponent to choose between rebuilding the damaged or destroyed asset versus investing in new assets, weapons, or intelligence gathering.[6] Thus, early attacks can be seen as an attempt to prevent an opponent from investing in capabilities that would threaten a player's future position.

---

[4] This latter analysis required that we define a significant attack. Because we did not want our results to be obscured by tit-for-tat attacks that characterize games of deterrence, we defined *significant* as a 10-percent drop in the opposing player's ability to project military power from space.

[5] In fact, for some time, we suspected that there might be a bug in our implementation of the mathematics of the game that was causing this to occur. Detailed examinations of the mathematics and the move-by-move scoring revealed this not to be the case.

[6] We model opportunity cost in our game by restricting the number of simultaneous investments a player can make. Therefore, the *cost* of an investment to a player is proportional to the investments that must be foregone during the time needed for a current investment to pay out.

Having established that strategic interaction patterns can be observed and characterized by looking at the overall pattern of the timing and intensity of attacks, we then found two more distinct clusters. In one cluster, both players reserve their attacks until just before or simultaneous with the start of a ground war. Furthermore, the attacks from both sides are focused primarily on reducing their opponent's attack capability. We termed these games "wars of weapons attrition." Neither side is interested in attacking space assets and incurring political stigma of having engaged in illegitimate acts. However, both used the outbreak of the ground war as a reason to quickly destroy their opponent's anti-satellite weapons—a legitimate act for which the game imposes no political penalties. Finally, we found that our remaining games exhibited a classic *horizontal escalation* of the ground war into the space domain. In these games, attacks are made well after the start of the terrestrial conflict and focus on creating exploitable shortfalls in the adversary's ability to project power from space.

In summary, by focusing on the timing and intensity of attacks, we were able to distinguish three distinct patterns of strategic interactions into which all games, other than games of deterrence or total destruction, fell. We believe these games provide valid insights into the dynamics we might see in future space conflicts.

Identifying Conditions That Result in Specific Strategic Interaction Patterns

Our goal is to develop an understanding of how the United States might leverage insights about these strategic interaction patterns to shape policy or future investments. Specifically, we want to answer the following research questions:

1. Can we identify the conditions that give rise to specific attack strategies?
2. Can investments in specific offensive or defensive capabilities be used to shape an opponent's attack strategy?
3. Can revelation or concealment of investments in offensive or defensive capabilities be used to shape an opponent's attack strategy?

Summary Answers and Recommendations for Future Studies

Although we found distinct patterns of strategic interaction, we were unable to construct a set of games that would allow us to "correlate" or even "strongly suggest" the conditions that give rise to those strategic patterns. Finding those correlations is key to developing an understanding of the types of investments that might allow us to achieve those conditions and hence "shape" a strategic interaction into a recognized pattern.[7] Despite strong evidence that investment and the conceal/reveal of investments have the potential to shape an adversary's attack strategies, clear rules about how to use investments to shape a future conflict elude us. We strongly suspect this is the result of one of the principles we found when exploring these questions: ***The dynamic value, both temporally and contextually, of an attack changes with the conditions of the game.***[8] In the next sections, we will elaborate on this principle and on possible implications for U.S. space policymakers.

For investments, variability that arises from temporal concerns is possibly easiest to understand. Timing of investments and attacks matters. In particular, we observe many incidences in our games in which the timing of an investment that would shore up a vulnerability or make obsolete an advantage creates a use-it-or-lose-it scenario for the adversary, precipitating early attacks that fit the early-shaping-of-the-battlefield strategy. We also observe instances where those early attacks are not being made solely to precipitate loss of advantage but to also divert investments into other realms such that key vulnerabilities or advantages are sustained. This latter consideration is a contextual concern. Diversion of resources to other objectives is only possible if there are diverse, but as yet unmet, objectives that a player

[7] For this reason, our subsequent research will move away from using large repetitive games with more than 400 possible moves in favor of simple game trees. Perhaps in doing so, we will develop insights that we can then bring back to the analysis of these larger games, similar to the insights we found when we looked for the pattern suggested by Sun Tzu's precepts regarding deception.

[8] This is a principle that space war shares with chess. See Appendix C for an extensive discussion of the principles of chess and go that may be applicable to warfare in the space domain.

values and there is scarcity of the resource. Both factors are dependent on game conditions.

It is harder to study the temporal and contextual circumstance around reveals of information. Overall, we found that a game-theoretic model is ill-suited to the study of the dynamics of reveal decisions. As a result, these moves are rarely used, and any attempt to find a pattern in their use is obscured by the randomness of the payouts.[9] We strongly urge that future researchers find other methods to explore the dynamics of reveals and deceptions. We believe that simple game trees can be constructed to provide insight into these issues, but we also believe that agent-based models (possibly each solving simple differential equations that specify behavior reminiscent of Richardson's work) can answer many of these questions. In addition to being able to model Bayesian updates of information that might give insight into the timing of reveals and deceptions, an agent-based model could incorporate surprise—an effect our game-theoretic models are incapable of modeling.

A Closer Look: Impact of Investments on Strategic Interaction Patterns

In our search to understand the conditions that determine which of the three strategic interaction patterns are used, we conducted a set of analyses to examine if there was a correlation between Blue's initial investment decisions and Red's attack decisions. Specifically, we wanted to understand whether Blue's initial decision to invest in redundancy and resilience of assets rather than offensive capabilities (i.e., weapons) changed the strategic interaction patterns. Details of our insights gained from these analyses are provided in the next sections, organized by strategic interaction pattern.

[9] The astute reader might observe at this point that we might have learned a great deal about the dynamics of hide/reveal had we run sets of games that varied the probabilities of belief versus bluff in a Bayesian manner. Indeed, the ability to compute Bayesian updates to each player's view of reality was an original goal of our game. We have discussed the computational burden that this imposes in other work. If we had infinite resources (or even faster computers than the supercomputers used), we undoubtedly would have pursued such an investigation. Instead, we have elected to revert to simpler game trees in forthcoming research to explore these issues further.

Early Shaping of the Battlefield

As discussed earlier in this report, when we set up a symmetric order-of-battle game with a small Blue vulnerability, Blue initially invests in redundancy and resilience to shore up that vulnerability. For instance, if we start a game where Blue's assets are vulnerable to EW attack and Red has EW weapons, Blue invests in hardening against EW attacks. Recognizing the diminishing future effectiveness of their weapons, Red responds by engaging in attacks early in the game to exploit Blue's vulnerability before those attacks become significantly less effective—a use-it-or-lose-it play. However, although this is one condition that leads to Red strategies of early shaping of the battlefield, we do not believe that this use-it-or-lose-it condition explains all instances in which we found this pattern.

In other more complex games, we observe that the early shaping can be equally well explained as an attempt by the opponent to force investments in redundancy and resilience at the expense of building weapons. For example, we used the same order of battle but instead of giving Blue an asset that is vulnerable to EW attack, we gave Red a small asymmetric weapons advantage. This had the expected effect of switching Blue's initial investments to building weapons in an attempt to achieve weapons parity.[10] Red's response in this case is slightly different. They engage in nondestructive but distracting reversible attacks against a wide range of Blue's assets, and there is more diversity in the weapon/target pairings used. In other words, Red's strategy is not simply an attempt to use EW effects against a particular asset before an advantage is lost. Red's attacks have the effect of switching Blue's investments to rebuilding redundancy and resilience for their assets. This is still an attempt to shape the battlefield early in the game, but instead of engaging in attacks to exploit Blue's diminishing vulnerability before it disappears, Red instead is seeking to maintain their own asymmetric weapons advantage.

It is worth taking a slight detour here to point out that when or if Red does decide to escalate to kinetic attacks in these early shaping games, this escalation takes place at or after ground war start. Red recognizes that the

[10] This is because, in times of peace, Blue values the differential in weapons capability. That is, their objectives value winning an arms race.

use of kinetic weapons before ground war start has an associated political cost that may not be worth it if Blue has sufficient defenses or resilience to mitigate the impact of an attack. If Red waits until ground war start to engage in kinetic attacks, they will not incur this political cost. Furthermore, attacks on assets prior to ground war start incurs political penalty for no military gain, but attacks on assets after ground war start provides military gain to offset the political penalties. Unless the Red player is in a use-it-or-lose-it scenario, they decide to wait to avoid paying the political cost.

This example illustrates how players can maximize the value of weapons use (or minimize the cost of weapons use) by making temporal decisions about when to use them. ***If a weapon will lose value if not used early in the game, players will not wait to use it. If a weapon will incur cost if used early, but not if used later in the game, players will wait to use it.*** This recognition of the dynamic value of weapons is a key insight from our research into the conditions that give rise to early shaping strategies.

Wars of Weapons Attrition

We believe that the dynamic value of weapons also explains much of the dynamics involved in wars of weapons attrition, but the reasoning is less straightforward. We believe the change in the value of weapons at ground war start is the primary impetus for the timing of the attacks.[11] Clearly, both sides have insufficient motivation that would lead to attacking space assets early—i.e., they have no need to shape the battlefield early—and it is the change in objectives that precipitates attacks. However, the dynamic value of weapons does not fully explain why those attacks are used primarily to attrit each other's weapons stockpile as opposed to each other's military assets. It must be that the value of destroying the adversary's ability to hold one's own assets at risk is more valuable than destroying the adversary's assets. Because we do not overtly weight these national objectives differently (preserving one's own power projection versus temporarily degrading the opponent's power projection), there must be some other temporal or contextual element in the game that creates this bias in targeting choice. There

[11] As we noted earlier, after the ground war start, the only valuation of a weapon is in what it can hold at risk.

are two aspects of the model and in reality that we believe lead to this bias in targeting:

1. The versatility of weapons. The fact that a weapon can hold *both* assets and weapons at risk may make them preferred targets in a war of attrition.[12]
2. The temporal effect of attacks on weapons versus attacks on assets. In our games, we model the loss of weapons as permanent. Once used or damaged, a weapon is removed from the game until a new investment to build weapons pays out. However, we model the degradation of power projection into the ground war as temporary. In other words, we modeled that workarounds to space power projection will always be found in the months or years following an attack. New investments may more rapidly rebuild space capabilities, but even without direct investments in space capabilities, we assume that nations will find workarounds—perhaps in terrestrial domains—to restore their military power projection.

We believe these modeling choices, although not a perfect representation of the real world, produce effects that are present in the real world. We discuss each in the subsequent list.

- **Weapons versatility**. Some anti-satellite weapons are designed to be most effective against a specific target—for instance, a jammer may be tuned to a specific frequency. But in general, satellites are more alike than different, and a weapon that works against one may have at least some limited effectiveness against all. Obviously, the more specialized the weapon, the less value it may have in terms of the diversity of tar-

[12] An astute reader might note that, given we modeled that each orbit can only absorb a set number of kinetic attacks before use of that orbit is destroyed, the number of kinetic attacks that can be prosecuted is a limited resource. When allocating these limited resources, small biases in valuing what can be held at risk may have large impact on the prioritization of targets. However, we found that wars of weapons attrition are not always conducted using kinetic attacks. Therefore, the bias in targeting is not solely dependent on scarcity of the number of kinetic attacks that can be made before destroying an orbit.

gets it can hold at risk. Conversely, the more specialized the weapon, the more effective it might be against a specific target. We model different effectivities for some weapon types against some asset types in our game and observe a definite preference for *using* general-purpose weapons over more highly specialized weapons. Whether this makes a more general-purpose weapon a more attractive target than a specialized weapon is dependent on context and thus more difficult to quantify. To date, we can make no general observation on this latter topic. However, our research highlights that *the diversity of targets that can be held at risk, not merely the effectiveness of specific weapon/target pairing, should be included as a factor in analysis of weapons design alternatives and in determining weapon/targets pairings for space battle management systems.*

- **Temporary impacts of degradations in power projection.** As just noted, we assume that nations will find workarounds—perhaps in other domains—to restore their military power projection if space assets are degraded. In fact, the United States is already preparing to do just that in the event that its Global Positioning System (GPS) is disrupted or degraded in future wars. In 2018, the U.S. Department of Defense issued Instruction 4650.08, *Positioning, Navigation, and Timing and Navigation Warfare*,[13] requiring all agencies to certify that their systems are resilient to interruption of any single source of positioning, navigation, and timing information, of which GPS is the most prevalent. This was a welcome and prudent development. Assets with shorter workaround periods are clearly less likely to be targeted in our games. *Publicly demonstrating that steps have been taken to lessen the dependency on space assets for U.S. military power projection is perhaps the most effective method to protect space assets that, like GPS, provide vital social, information, and infrastructure capabilities.*[14]

[13] Department of Defense Instruction 4650.08, *Positioning, Navigation, and Timing and Navigation Warfare*, Washington, D.C.: U.S. Department of Defense, December 27, 2018, change 1 effective December 30, 2020.

[14] If nations engage in wars of total destruction, any asset that provides capabilities in any of the domains of national power—including social, information, and infrastructure—will be targeted. However, in more limited war, we believe redundancy

Horizontal Escalation of Ground War into the Space Domain

The horizontal escalation of war into space after ground war start is the least frequent strategic interaction pattern found in our results. The small number of games in which this occurs makes it difficult to formulate hypotheses about the temporal and contextual factors that may give rise to these games. We can say with some amount of certainty is that it is not correlated to initial conditions of the game but is instead affected by the moves available to the players in the game. As we described in Chapter Two, players add moves to their move sets over a series of games, and often a game that results in early shaping or a war of weapons attrition is transformed into a game of horizontal escalation by the addition of new moves. We have found no pattern as to the type of moves that are added (i.e., whether they increase or decrease which capabilities can be held risk) that would explain this shift in strategies.

---

and reduced military reliance on social "goods," such as GPS, can be a powerful deterrent through denial of gains. There are simply more vulnerable and more valuable targets for an adversary to attack.

CHAPTER FIVE

Summary and Recommendations for Future Work

Strategic Principles of Conflict in Space

Here, we summarize key principles from our research about how a future conflict may play out in space. We include recommendations based on these principles where applicable.

Principle 1: In an offensive-rich environment, improving defenses against one attack vector may shift the adversary's tactics, but it is unlikely to shift their strategy.[1]

- Developing a comprehensive defense against kinetic attacks may shift adversaries away from kinetic attacks and toward less debris-generating attacks—a result that enhances society's ability to preserve the use of space for future generations.
- To effectively move U.S. adversaries away from kinetic attack and into other less debris-generating weapons may require public demonstration of U.S. defensive capabilities against kinetic attack.

---

[1] *Offense dominance* was an important concept in nuclear deterrence and refers to a scenario in which a conflict is heavily weighted in favor of the offense, that is, defenses are largely ineffective against a determined offense. For further discussion, see, for example, Forrest E. Morgan, *Deterrence and First-Strike Stability in Space: A Preliminary Assessment*, Santa Monica, Calif.: RAND Corporation, MG-916-AF, 2010; James P. Finch and Shawn Steene, "Finding Space in Deterrence: Toward a General Framework for 'Space Deterrence'," *Strategic Studies Quarterly*, Vol. 5, No. 4, Winter 2011.

Principle 2: There is no deterrence by defense in a conflict where deploying an offense comes with a relatively low cost to the attacker, while deploying a defense require a high price from the defender.

- It is not enough to simply build systems that enable defensive actions—nations must also invest in ways to reduce the cost of employing the defense if they are to deter attacks.
- The cost of taking a satellite out of service during a ground conflict—i.e., the certainty of incurring a self-inflicted wound—may rarely be worth it when compared with the less certain probability of kill.

Principle 3: Decisions on whether to reveal or conceal investments in, or the actual extent of, offensive capabilities are not straightforward, especially when considering that not all potential opponents in space are peer competitors. Although our research using the game finds a slight bias toward revealing *defensive* capabilities, we find a strong bias against revealing *offensive* capabilities.

- Misperception regarding offensive capabilities can be stabilizing when (1) the adversary perceives a peer opponent to be disadvantaged or (2) a disadvantaged opponent is perceived to be a peer.[2] In other conditions, we found misperception of offensive capabilities to be destabilizing or to have no clear trend.
- The United States has the ability to shape adversary decisionmaking through investments and possibly what it chooses to reveal or conceal about those investments. However, the optimal message depends on the adversary, and the benefit of such messaging will rely on understanding the adversary and the ability to control this message for each potential adversary.
- Regarding whether the United States should make greater efforts to clarify the intent of GSSAP and the X37-B, uncertainty appears to be generally stabilizing if the concern is a future conflict with a near-

2 We say that an effect is stabilizing if the resultant conflict is of lower intensity. Our metric for conflict intensity takes into account the perceived escalatory nature of attacks as well as the character and number of attacks.

peer adversary. If the concern is instead a conflict with a less capable adversary, we believe from our prior work that other factors, such as an asymmetric U.S. dependency on space to project power into ground wars, will dominate the determination of whether the United States can deter the escalation of that conflict into the space domain.[3] Revealing or concealing offensive capabilities may make no difference. Therefore, overall, we believe the United States should continue to operate these satellites in a nonthreatening manner to avoid creating a use-it-or-lose-it scenario for possible adversaries. We also believe that it is not necessary, and may even be detrimental, to remove all ambiguity about U.S. capabilities.
- Publicly demonstrating that the United States has taken steps to lessen the dependency on space assets for military power projection is perhaps the most effective method to protect space assets that—like GPS—provide vital social, informational, and infrastructure capabilities.[4]

Principle 4: In a strategic competition in space, the value of any given investment or attack is heavily time-dependent, context-dependent, and dynamic.

- If a weapon will lose value if it is not used early in the game, players will not wait to use it. If a weapon will incur cost if it used early, but not if it used later in the game, players will wait to use it.
- The fact that a weapon can hold *both* assets and weapons at risk may make them preferred targets in a war of attrition.
- The diversity of targets that can be held at risk, not merely the effectiveness of specific weapon and target pairing, should be included as

[3] Triezenberg, 2017, varies dependency on space to illustrate that it may be impossible to deter a player who is less dependent on space from horizontally escalating war into the space domain.

[4] If nations engage in wars of total destruction, any asset that provides capabilities in any of the domains of national power—including social, information, and infrastructure—will be targeted. However, in more limited war, we believe redundancy and reduced military reliance on societal "goods," such as GPS, can be a powerful deterrent through denial of gains. There are simply more vulnerable and more valuable targets for an adversary to attack.

a factor in analysis of weapons design alternatives and in determining weapon and target pairings for space battle management systems.

- Game-changing moves often are *not* played. Instead, merely the possession of the move changes the decision calculus for using other moves: It is the *potential* for these moves to be played that changes the game. This potential can act as a deterrent to the opponent or a reassurant to the possessor. That is, sometimes what changes the game is that possession of a move allows a more capable player to bide their time and not escalate a situation they might have otherwise felt compelled to contest.
- Although players prefer tactics that are effective over wider ranges of conditions, if a widely effective tactic is denied, they will find other ways to accomplish their goals. It is the plentitude of tactics available in the space domain that may limit U.S. ability to affect an opponent's strategic goals. We found no silver bullets.

Finally, we note that if space is to be a factor in the outcome of future wars against near-peer competitors, our work leads us to believe that it will be won at the margins. Space will matter only if a side can better exploit temporary advantage.

Strategy Characterization and Observations

The overarching question throughout this research is how U.S. investment decisions today will affect strategic interaction in a future space conflict. All the games we ran shed light on some aspect of this question, and what we have gained from our years of game analysis is an appreciation for the distinction between strategic and tactical decisionmaking. Although tactical decisionmaking identifies specific actions that a player should take to achieve optimized outcomes in the short term, strategic decisionmaking looks at the effect of actions and investments over the course of the game or conflict. Our game model is not intended to provide guidance for tactical decisionmaking, and indeed there is a probabilistic element to player decisionmaking that results in randomization in the set of decisions made by each player even for identical games. We believe the value of our game is in examining broader player strategies and how investments or mindsets can alter game outcomes from this larger strategic view.

In our research, we identify levers that may shape the timing and cadence of a conflict. We find patterns of strategic interaction that can be characterized by their timing with respect to the start of a ground war.[5] We term the first *early shaping of the battlefield.* In these games, attacks occur two or more years before ground conflict in what may be an attempt to shift the opponent away from investing in weapons and instead invest in recovering the space capabilities lost in the attack.[6] We also find that early shaping of the battlefield can be the result of creating conditions in which a player may lose the effectiveness of a weapon if they were to delay its use until the start of the ground conflict (i.e., a use-it-or-lose-it condition has been created). We characterize the second strategic interaction pattern as *wars of weapons attrition.* In these games, both sides reserve their attacks until just before or simultaneous with the start of a ground war. These attacks from both sides focus primarily on reducing an opponent's attack capability. We hypothesize that these games reflect conditions in which both sides have no military incentive to attack assets that incur political penalties and so reserve their attacks for legitimate targets (i.e., weapons) at ground war start. Finally, our third strategic interaction pattern is observed to be a classic *horizontal escalation* of the ground war into the space domain. In these games, attacks are made well after the start of the terrestrial conflict and focus on creating exploitable degradations in the adversary's ability to project power from space.

These patterns are related to the start of the ground war because we changed our players' objectives at that point. We did this to mimic real life: In a ground conflict, the ability to exploit vulnerabilities and create short-

[5] To clarify, these strategies are in addition to those in which both sides maintain the status quo (deterrence) and engage in a war of destruction. Although those games are interesting, prior publications have dealt with the conditions that we found give rise to them. In this report, we focus on conflict dynamics short of total war.

[6] We also found a pattern of early shaping of the battlefield that we term *brinksmanship.* It is unclear whether this pattern is simply a manifestation of known issues with game-theoretic models or if it represents a pattern we might expect to find in the real world. The brinksmanship pattern is found in many of our games in which both sides engage in a war of destruction. We describe this in Appendix C.

falls in an opponent's power projection into the ground theater increases in importance. Furthermore, in the case of war far from the homeland, the ability to project power through space may become even more critical. Additionally, in times of war, opponents may care less about perceived status in the international community in the long term than their own self-preservation in the short term. All games reported here begin during peace time, with players seeking to maximize their capacity to project military from space, political standing in the world, and weapons capabilities relative to their opponent (i.e., winning the space arms race). The length of this period varies based on the objective of our research—whether it is focused on the investments that lead up to the start of a ground war or on the dynamics post–ground war start. But, in all games reported here, it is set between five and seven years into the game. At ground war start, we change the objective: Players no longer seek to maximize their relative weapons advantage and begin to value minimizing their opponent's ability to project power from space into the ground war. In other words, the space arms race ends, and the space power projection conflict begins. Deterrence is not an explicit objective of either player—deterrence only arises if no player can improve their standing by engaging in attacks. At the same time, space war is not an objective: Players only attack if doing so will improve their overall score at the end of the game and will often attack before ground war start if that is in their best long-term interest. As a result of this shift in objectives, it is unsurprising that we found patterns of strategic interaction that are defined relative to ground war start.

Real-World Applications

Thucydides Trap predicts conflict will arise between an established power and a rising power because of "structural stress."[7] It is the changing balance of power between two nations that gives rise to conflict. Having been one of

[7] Graham Allison, *Destined for War: Can America and China Escape Thucydides Trap?* Boston: Houghton Mifflin Harcourt, 2017, provides an overview of Thucydides Trap and its application to historical and current great-power competition. Allison postulates that these structural stresses present between the U.S. and China could increase the likelihood for conflict.

two dominant powers in space for many decades (the former Soviet Union was the other), the United States is clearly an established space power. However, investment in both military and commercial space capabilities are on the rise in many nations. Some of those nations are approaching, or hoping to soon approach, peer status with the United States. Therefore, we offer two observations from the game that may provide insight into how the investments of a rising space power may play into decisionmaking on either side.

One way by which investments may create structural stress between nations is by creating a use-it-or-lose-it scenario. We clearly observe cases in our games in which the established power reacts with a strike-first strategy to investments by a rising power to improve its resilience and reduce the effectiveness of the established power's weapons capability. We observe that the net result of these early strikes is often a war of total destruction in space. We hypothesize that this is because the rising power cannot achieve the necessary redundancy and resilience needed to achieve deterrence prior to the start to the ground war. Timing matters.

Another incentive to strike first may be entwined with peacetime objectives to win the space arms race while maintaining power projection from space. If a rising power begins the game with a deficit in weapons but starts to develop weapon capabilities, the highest payoff for the established power is to prevent a rising power from doing so and thus preserve relative dominance in the space arms race. We observe several cases in which an investment by a party to build weapons appears to precipitate an adversary's early attacks on assets. The adversary is attempting to distract the party developing weapons and instead divert their resources into rebuilding redundancy and resilience for the damaged assets. One of the reasons we hypothesize that this distraction is such an effective strategy for the adversary is that redundancy and resilience are investments most highly correlated to achieving deterrence of escalating future conflicts into the space domain.[8] In this case, the short-term desire to compete in the arms race is overcome by the longer-term need to maintain the ability to project power through space in a future terrestrial conflict. A well-timed first strike in these cases rarely leads

[8] See Triezenberg, 2017, for an exposition on why redundancy and resilience investments are so highly leveraged in deterrence.

to wars of total destruction. The key is giving the rising power enough time to achieve resilience prior to the start of ground conflict.

However, not all structural stress factors arise because a player is attempting to stymie another's rise. Sometimes, strike first occurs simply because a player perceives a higher gain for conflict than for the status quo. We often see one or both players willing to engage in actions that will result in the mutual loss of use of space, a conflict that would not be a priori preferable to either side. In a quest to maximize shortfalls in an opponent's ability to project power from space while minimizing their own, players will elect to destroy the use of space for all if it locks in an advantage in the ground war. In prior work using the model, we show that a U.S asymmetric dependence on space is a strong motivator for less-capable and less space-dependent adversaries to adopt a strategy of total destruction.[9]

Recommendations for Future Work

This report provides an overview of our observations about conflict dynamics and the conditions that accompany them. Although we can develop hypotheses about the causal relationships between different conflict shapes and outcomes, we were unable to fully determine or test them. This limitation was a result of the computational complexity of the model. Although we caution against the use of any game-theoretic model as a predictive tool, developing a better sense of the causal dynamics at play could help build a more informative narrative about the forces at play in a conflict. Future work using a similar game-theoretic approach could benefit from a targeted examination of more bounded questions to elucidate some of these cause-and-effect conditions.

The results of several analyses here point toward the utility of deeper study. We observed the result that SDA-enhanced defenses lead to a change in adversary tactics but not overall strategy. However, we also observe that certain circumstances could lead to more favorable tactics. More targeted studies of strategic interactions in space could help identify these favorable tactics even as long-term adversary strategies remain unchanged.

[9] See Triezenberg, 2017.

The nuanced results from our hide/reveal analysis motivate a more extensive examination of messaging, trust, and transparency. In this report, we discuss a model that does not accommodate a consideration of the level of uncertainty about an opponent's capabilities. When Blue hides, Red does not have uncertainty about its own lack of information. It simply makes decisions assuming that its misperception is truth. Additional work could leverage a similar but simpler game model, along with existing work that RAND and others have done on strategic messaging, to determine how uncertainty about perceptions could impact conflict outcomes and strategic interaction patterns.

APPENDIX A

Overview of Project Phases

The project outlined in this report developed a game-theoretic model of space security to represent qualitative theories of state competition and human behavior in the quantitative rules of the game. The outcome of this work is a complex game that examines the effect of space investments in offense, defense, and intelligence gathering on future space warfare. The game uses prospect theory–derived deviations from rationality, misperception, deception, escalation, and payoff uncertainty to determine effect on future strategies. This software provides operational insights into alternative futures to support decisionmaking by the sponsor.

In Phases I and II of this work, which took place from 2015 to 2016, we built the game mechanics and software, laying the groundwork for more-nuanced explorations. Torrington et al. (2019) contains the final report for these phases of the contract. Under the Phase III and Phase III extension contracts (2017 and 2018), we added game rules to better explore effects of U.S. investments in SDA. We also improved representation of signaling among game actors. Signaling capabilities now include the ability to hide or reveal actor objectives (redlines and deadlines) and capabilities (misperception and deception). During Phase III, we also collaborated with Lawrence Livermore National Laboratory and the National Reconnaissance Office's Centralized Super Computing Facility to transition the game to a supercomputing architecture, allowing us to more fully explore issues of interest to the sponsor. In Phase IV, we examined specific strategies and game-changing scenarios and explored factors that may impact strategy selection among players. The work summarized in this unclassified report was primarily conducted in Phases III and IV.

TABLE A.1

Summary of Phases I–IV of Project

Phase	Objectives
Phase I	Determine core mechanics of the game.
	Incorporate prospect theory into game analytics.
Phase II	Develop software for solving the game and refinement of game mechanics.
Phase III	Implement variable time horizons to model short-term versus long-term planning.
	Transition game from a PC to supercomputing platform.
Phase IIIx	Explore game scenario designed for targeted decisionmaking support with varying Red and Blue orders of battle, details of which are documented in reports at higher classification levels.
	Explore misperceptions of both offensive and defensive capabilities.
	Explore ways to leverage the game to support future human war games.
Phase IV	Analyze specific game-changing scenarios.
	Use research results and products to participate in and support human wargames.
	Synthesize lessons learned to develop and document a framework for understanding how future U.S. investments are likely to stimulate adversary actions in the space domain.

APPENDIX B

Game Structure and Methodology

Our game is a two-player game-theoretic model developed to help build insight into adversary behavior, strategy, and outcomes in a space conflict. We designed this model to allow us to explore a variety of potential levers over an extensive time horizon, examine how strategies evolve over time, and explore decisionmaking from various types of players and sentiments.

Representation of Adversary Decisionmaking

Game theory is a tool to build an understanding of adversary decisionmaking. Thus, to provide valid insights, a game-theoretic model must include realistic representations of how an adversary perceives actions and the context of the conflict, how they reason, and how they make decisions. A key part of abstracting real-life conflict and its associated actions is to understand the decisionmaking calculus of the opponent based on their perceptions. Capturing the richness and complexity of players is vital to the development of a realistic model of decisionmaking. Additionally, to build a rich representation of these perceptions, the game builds on traditional game-theoretic representations of opponents in four ways:

1. Players may deviate from strictly rational valuation of outcomes.
2. Players may have multiple and asymmetrically weighted objectives.
3. Players may have imperfect information about adversary capabilities.
4. Players make decisions based on a finite look-ahead.

Prospect Theory and Player Payoffs

The first method of representing adversary perception is by using prospect theory to realize the skewing of the perception of risk and reward based on the reference point of the adversary player. Prospect theory helps us address one of the unrealistic aspects of rational choice theory. This is the assumption that people make decisions based on mathematical reasoning about the probabilities of success or failure in evaluating courses of action. Yet work in psychology demonstrates that this is almost never the case. For example, when stakes are high, emotions may color our weightings of gains and losses. Kahneman and Tversky (1979) articulated a theory of how people evaluate probable gains and losses, and they and others who subsequently developed prospect theory also built a corresponding mathematical model to represent this behavior. The model is based on experimental data reflecting our human inability to correctly evaluate probabilities based on the psychological effects of gains or losses associated with a course of action.[1] A graphical representation of how payoff values are skewed with the application of prospect theory is presented in Figure B.1.

Multi-Objective Players

We add further richness to our characterization of the adversary by representing them as non-unitary actors. Each player in our game has independent sentiments regarding different dimensions of national interest. Borrowing from an approach used by U.S. military strategists to examine critical dimensions of state power—namely, political, economic, and social dimensions, as well as those involving military, infrastructure, and information[2]—we considered separately how moves in space affect these dimensions, adding significant nuance to the discussion of how sentiments affect deterrence.

For each game run, a set of equations representing each of these objectives is solved. This determines the moves that maximize each player's objec-

---

[1] Kahneman and Tversky, 1979. It is important to note that players in our game may be nonrational but are not irrational; that is, they still will choose the option that provides the greatest payoff *as they perceive it.*

[2] Joint Publication 3-0, 2017.

FIGURE B.1
Perceived Payoff Values Represented in a Prospect Curve

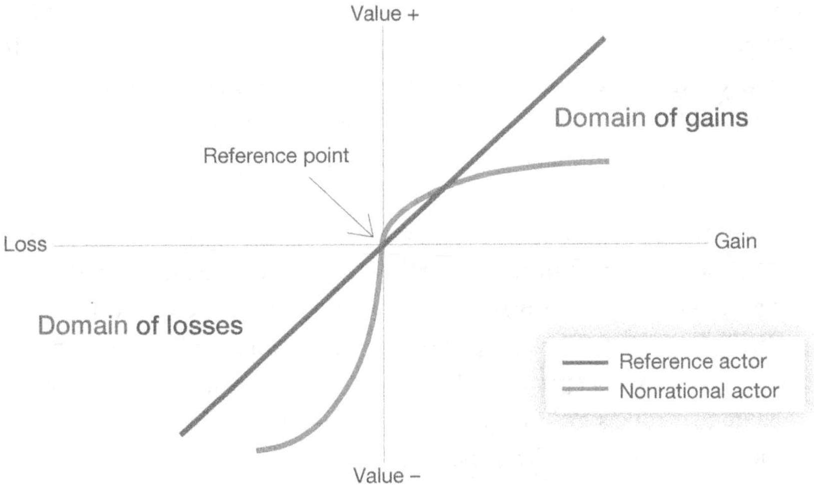

SOURCE: Based on our analysis of Kahneman and Tversky, 1979.

tive while the opponent simultaneously maximizes their own. This is not necessarily zero-sum, because some objectives, such as the ability to project military power from space, are valued independent of how the adversary scores in this objective.

To track game dynamics and results, we assess a score for each objective for each player, which means that our analysis is inherently multidimensional. For example, an attack against the GPS used to navigate U.S. warships and guide precision bombs pays off militarily for an adversary, but, because that same satellite system is also essential to the everyday business and social transactions of billions of global users, the attack also creates a loss of SII that could prove escalatory. Furthermore, the GPS satellites carry a sensor for detecting nuclear events, and an attack could be misconstrued as preparation for nuclear war, politically stigmatizing the attacker while creating sympathy for the United States. The net payout would be a significant strengthening of U.S. political power on the international stage. All players, therefore, must weigh choices that could provide benefits in one aspect against losses in another.

Imperfect Information About Adversary Capabilities with Finite Look-Ahead

The game was built to help understand how conflict or cooperation in space might play out over an extended time horizon. Players evaluate each branch of the game tree to the end of the game (or as far ahead as we allow) and then use backward induction to choose the optimal move at each node. However, not only is this process computationally large, it is also unrealistic. Rarely are humans able to accurately evaluate how decisions may play out hundreds or even tens of moves ahead. For the ten-year time horizon of the games described in this report, we have moves available daily.[3] To more accurately represent player decisionmaking in this context, we place a six-move limit that the player can see forward from their current decision node.[4] This means that players will optimize their score at each node for the sub-game that consists of the next six moves, but the overall game score for each objective represents the game in its entirety.

Game Structure

This game is designed to represent long-term strategic planning for two players rather than a simple move and response. When solving the game, the sequence of moves selected by each player optimizes their objectives subject to the selection of moves that optimize the opponent's objectives.

[3] The daily time scale was used to account for some of the shorter payoff and attack impacts. For example, a deployed defense will cause a service outage of days, not weeks. Similarly, attacks to deny, deceive, disrupt, and degrade—i.e., any intent other than the destruction of the target—only have an effect of days. To mitigate the computation burden of the time step, the solver skips ahead; if no move pays out, no effect terminates on a given day, or, if no other threshold is breached that would change the player's objective functions, that day is not evaluated. Nothing has changed that would make a player's decision different than that of the day before.

[4] The limit of six was chosen as a reasonable representation of how far ahead most humans think about how decisions play out. Although the best chess players can think ahead further than this, thinking through the hundreds of possibilities in a space war is more daunting than thinking through the more limited options available on a chess board. It is important to note that the limit is *not* six days. Although players have moves available each day, on most days, they elect to do nothing because nothing has changed.

Equilibrium solutions therefore represent the best possible outcome for each player—no other sequence could make one player better off without making the other player worse off.

The model is represented as an *extensive-form game* (see example in Figure B.2), in which each set of action choices by a player leads to an explicit new state of the game and a new decision point. This process continues until an end state is reached. Using the illustration shown in Figure B.2 as an example, an extensive-form game can represent the situation in which a player at decision point P_1 can either take the set of actions represented by A or the set of actions represented by B. The set of actions the players take dictate which branch of the tree the game proceeds along until a new decision point is faced. The bottom of this graph represents terminal states where each player receives some final payoff. For example, a player at point P_1 can chose to invest in a new capability (path A) or can elect to do nothing (path B). Later, when the investment has payed out (point P_2), their opponent can elect whether (or not) to attack that new capability. If the decision is made to attack (path C), then the first player must decide whether to absorb the

FIGURE B.2

An Extensive-Form Game

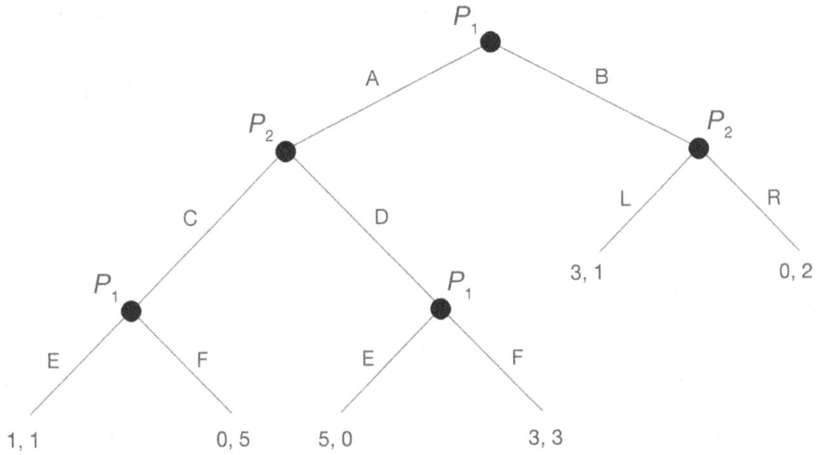

SOURCE: Based on the description of extensive-form games outlined in John Von Neumann and Oskar Morgenstern, *Theory of Games and Economic Behavior*, Princeton, N.J.: Princeton University Press, 1944.

attack (path E) or to defend against that attack (path F), which may be a lower overall payout for them and a gain for the attacker if the defense takes an asset out of service. Clearly, the payouts here—a (1, 1) score if no defense is made and a (0, 5) score if the defense is not activated—would be a better result for the first player.

Our game-theoretic model is much more complex than the figure's. Rather than playing one or two moves, our game-theoretic model plays out over multiple years, during which time the players have moves available every day. Furthermore, a player's choices are not binary. Indeed, they can make multiple choices each day, including whether to (1) invest in improving assets, defenses, or intelligence about their opponent; (2) attack a specific asset with a specific weapon for a specific purpose; and/or (3) defend a specific asset using a specific defense.[5] As we noted earlier, at each time step in our game where a player is allowed to make a decision, the full set of possible moves that could be made is usually around 200, depending on the game setup. The number of moves is a function of the number of assets and weapons each can invest in, types of investments, unique weapons/target pairings and the intent of attacks, and possible defenses that could be invoked. Furthermore, each of these decisions has a different payout, depending on the current state of the game. We also allow for imperfect information in the model about initial player capabilities, as discussed in the analysis about hiding and revealing weapons. This results in needing to solve three different views of the game—Blue's, Red's, and the actual game truth. Blue's view includes their own capabilities, their perception of Red's capabilities, and their perceptions of what Red knows about Blue's capabilities. Red's view is the opposite. Figure B.3 depicts these different views of

[5] Although the mathematics of the game are structured to allow for multiple simultaneous investments and simultaneous attacks, all games in this report were constrained such that a player's investment or attack must pay out before they can initiate a new investment or attack. This way, time is used as a proxy for the opportunity cost of making investment or attack decisions. This is not unrealistic. Investments or attacks in space that have long durations before paying out are often more costly in terms of resources (money, time, effort) than those of shorter duration. Per Merriam-Webster, *opportunity cost* is "the added cost of using resources (as for production or speculative investment) that is the difference between the actual value resulting from such use and that of an alternative (such as another use of the same resources or an investment of equal risk but greater return)."

the game. Given this complexity, estimates of the number of terminal nodes to be evaluated quickly approach the trillion mark for any two reasonably space-capable opponents.[6]

Objectives in the game are different based on the epoch of the game. *Epochs* are set by either the passage of time (a *deadline*) or the breach of a red line. This allows us to simulate the phases of conflict. All games in this report begin during peacetime with players seeking to maximize their capacity to project military from space, political standing in the world, and weapons capabilities relative to their opponent (i.e., winning the space arms race). The length of this period is variable depending on whether the objective of our research is focused on the investments that lead up to the start of a ground war or on the dynamics post–ground war start, but in all games reported here, it is set between five and seven years into the game. At ground war start, we change the objective. Players no longer seek to maximize their relative weapons advantage and begin to value minimizing their opponent's ability to project power from space into the ground war. In other words, the space arms race ends, and the space power-projection conflict begins. Deterrence is not an explicit objective of either player. It only arises if both players cannot improve their standing by engaging in attacks. At the same time, space war is not an objective. Players only attack if doing so will improve their overall score at the end of the game and will often attack before ground war start if that is in their best long-term interest. All objectives are structured mathematically as integrals of time. This means that the effect of a loss of 10 percent of military capability for one day is equal to the loss of 1 percent of military capability for ten days. The mathematical equations solved are documented in Triezenberg (2017) but are repeated in the numbered list for reference. Detailed rationales for why the math is structured as it is are provided in Triezenberg (2017) and are not repeated here.

---

6 In Appendix B of Torrington et al., 2019, the game-tree complexity of our model is estimated to be 10^{240} for a game with 20 weapons and 20 assets if we were to allow three concurrent investments and two concurrent attacks. This computation is only for game truth. Introducing misperceptions means we have to solve the tree twice more: once for Blue's view of truth and again for Red's view of truth.

FIGURE B.3
Players Optimize Actions Based on Their View of Game Truth

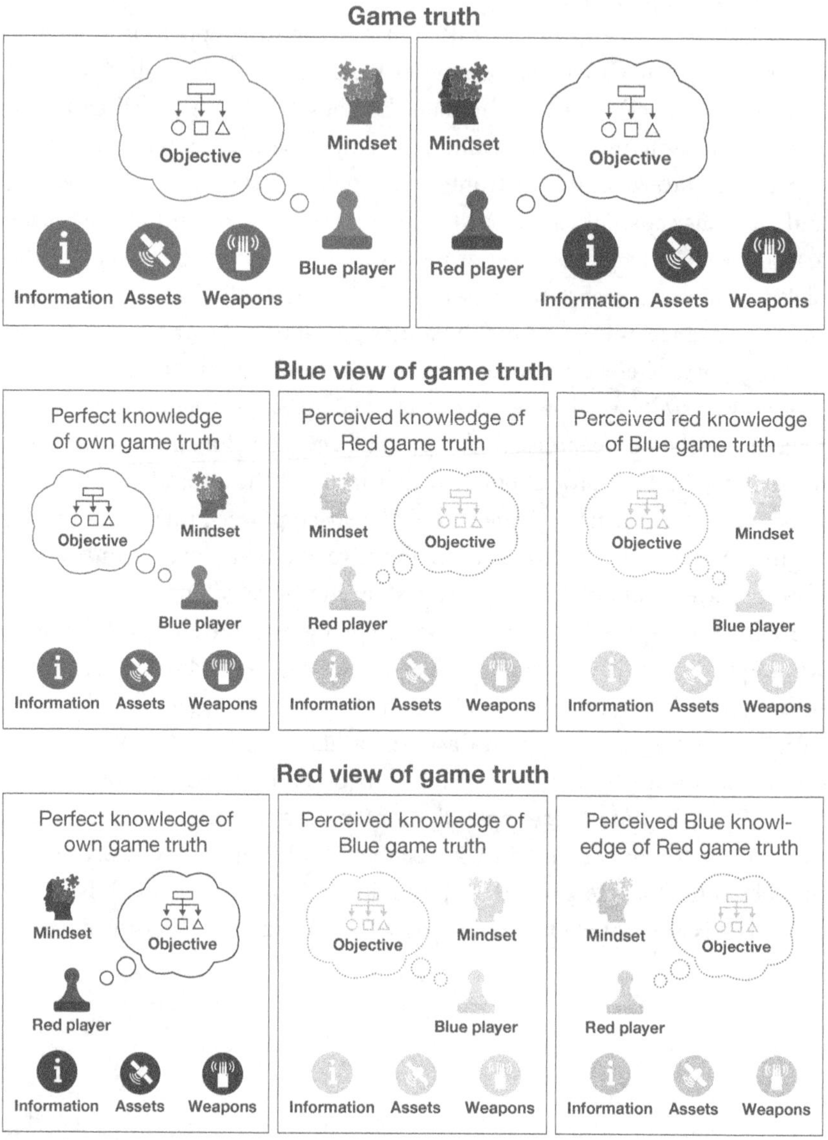

SOURCE: Adapted from Torrington et al., 2019.

1. Players seek to maximize the integral of the *M* capability of their own space assets over time (t, where T_G = game duration).

$$M_T = \sum_{t=0}^{T_G} (M) \times \Delta t$$

2. Players seek to maximize the integral of the *SII* capability of their own space assets over time.

$$SII_T = \sum_{t=0}^{T_G} (SII) \times \Delta t$$

3. Players seek to maximize the integral of their opponent's military (*Opp M*, where $_{Tbt}$ = time below threshold) shortfalls; [i.e., *M* capability below threshold] over time (t, where $T_{Gnd\ War}$= duration of the ground war).

$$Opp\ M_{Tbt} = (T_{GndWar}) - \sum_{t=T_G - T_{GndWar}}^{T_G} (Opp\ M) \times \Delta t$$

4. Players seek to maximize the integral of their opponent's SII (*Opp SII*) shortfalls (*SII* capability below threshold) over time:

$$Opp\ SII_{Tbt} = (T_{GndWar}) - \sum_{t=T_G - T_{GndWar}}^{T_G} (Opp\ SII) \times \Delta t$$

5. Players seek to maximize the integral of their own P capability (P_{LIM}) over time.

$$P_T = \sum_{t=0}^{T_G} (P_{LIM}) \times \Delta t$$

where

$$P_{LIM} = \begin{cases} 1, \textit{if } P > 100\% \\ P, \textit{if } 0\% \leq P \leq 100\% \\ 0, \textit{if } P < 0\% \end{cases}$$

6. Players seek to maximize the integral of the delta between the M capability of their own space weapons over time and the opponent's over time:

$$\Delta M_{wT} = \sum_{t=0}^{T_G} (Own\ M_w - Opp\ M_w) \times \Delta t$$

Solving the Game

Although previous reports have captured the game mathematics in more detail,[7] this report concentrates on describing the concepts and methods behind decisionmaking in the game and what this can tell us about the stra-

7 See Torrington et al., 2019, and Triezenberg, 2017.

tegic interaction overall. Importantly, decisionmaking in the game is based on how the player perceives the state of the game. The perceptions of the players as described earlier in Figure B.3 are used when determining what actions they will take, not the game truth. Although we track game truth to determine objective outcomes throughout the game and make any policy recommendations, each player's reasoning process is based on their individual perception.

The game model used for our analysis is necessarily complex, given that attacks and defenses of space assets are constrained by the dynamics of maneuvering in space, capabilities of the weapons used, defenses of the assets attacked, and myriad ways in which both offensive and defensive means can be employed. Each player's cognitive-decision model is defined by both an *objective* and a *mindset*. Play is called *rational* if objectives are evaluated based on a player's observable game truth. Play is termed *prospect* if objectives are evaluated based on subjective evaluations of gains and losses shaped by the player's sentiment as postulated by prospect theory. In all of our research, we play identical games under rational choice theory and under prospect theory to ensure that any recommendations we might make are valid under both theories of decisionmaking.

As noted earlier, the number of terminal nodes in the game tree approaches the trillion mark rather quickly for any reasonably complete simulation of real-world space adversaries. This led Triezenberg (2017) to constrain the games for her dissertation to toy problems where each player had only three asset types and three weapons types, but we knew another solution would have to be found if we were to conduct meaningful research for our sponsor. Initially, we had hoped to divide the game tree itself into parallel paths for computation, but even that was quickly found to be beyond the capabilities of the computing infrastructure we had available to us.[8] The solution lay in realizing that although the number of possible moves in the

[8] It was at this point in the development of the game that we teamed with Lawrence Livermore National Laboratory to improve the computation times of the game. A quick calculation by one researcher at the laboratory predicted that computing all of the nodes for the game tree would take decades to solve even one game. In fact, a brute force solver of a three-asset/three-weapon game was begun in 2016 and was still running in 2019 when the laptop it was executing on died from a hard drive failure. In that time, it had computed nearly a billion nodes.

game was large, the actual number of moves that appeared in solved games was very small (mathematically, there are few pure strategies). This led us to structure the game such that an initial game with just a few a priori selected moves for each player is solved first and then each player adds a move to their move set to determine if doing so improves their outcomes. It is the testing of each additional move that is parallelized on the supercomputer. The two moves that most improve the game score are then added to the game, and the game is solved. Thus, the game is solved iteratively with each improved move set for a player counted as an iteration.

As discussed in Chapter Two, for the initial iteration (termed *iteration 0*), each player selects a set of moves for play. Those moves are selected to maximize the players' objectives but without considering that the adversary is also making moves to maximize their objectives. Iteration 1 is the first solved game in which, at each time step, players determine what moves to make (from among those in their move set) to maximize their score, but now subject to the opponent also making moves to maximize their own objectives. This process continues with each player taking turns adding moves to their move set until the game results converge and neither player can improve their score by adding moves (i.e., we have found a local optimum) or until the size of the games becomes so large that solving for the effect of the added move takes too long to be of value to our research (generally, we stopped adding moves to the game when the process of checking for an improved score exceeded one week—usually after eight to ten iterations). Figure B.4 summarizes this iterative process pictorially.

FIGURE B.4
Building of Move Sets During One Game Iteration

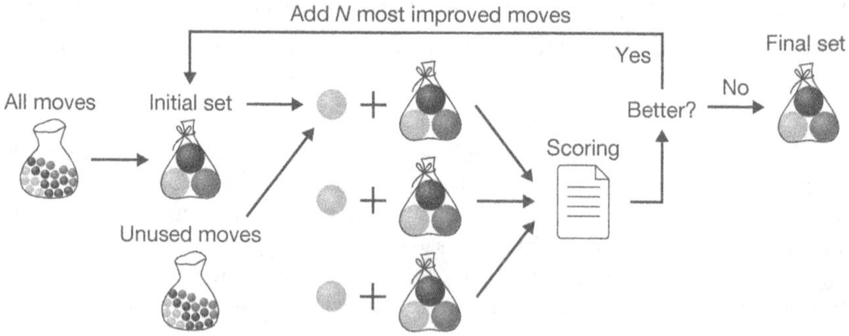

It should be noted that there still may be combinations of moves that would improve the game scores, so even if we were to continue adding moves until game results converge, the move set is not, strictly speaking, optimal. It is, however, as close to optimal as we can get within the computational limits of today's technology and a reasonable time limit. As more moves are added to the games, they become more and more difficult to solve. A game with set of 15 to 20 moves, for instance, may take weeks to obtain results. We have come, however, to believe that optimality in the move set is not necessary for our work. Our research questions center on how investments shape the range of behaviors that could be expected in future space wars and for this reason we view all games after each player has had a chance to add moves to their move set (i.e., above game 3) as valid sources of data on which to base observations.

Managing the Model Configuration

The model is written in Java, and the game runs used for the analysis discussed in this report were conducted using sponsor agency supercomputers. Torrington et al. (2019) include a description of how the game was validated using a combination of inspection of the code base by independent analysts and select test cases. In addition to the model validation described in that report, the source code and documentation were subjected to an independent quality assessment in 2018. A regression test case that exercises nearly 80 percent of the paths in the code is run at each major upgrade of the code base, and the code itself is maintained in a GitHub repository at RAND. For each major upgrade, RAND delivers a copy of the source code and user manual to the sponsor. The U.S. government has the right to make a derivative work as required by the Federal Acquisition Regulation. A complete inventory of all runs made, including input files used and output files generated is maintained on the sponsor's computing system. The inventory also contains version-controlled Excel-based macros that are used to compute metrics and conduct analyses, as well as the output of the macros.

APPENDIX C

Lessons Learned from Using Artificial Intelligence to Conduct Research

In nearly all of our work with the model, an iterative approach to human-machine learning was needed. For those interested in what is called *hybrid learning*, this appendix describes lessons from the journey. First, we discuss lessons learned in developing the appropriate abstraction for the model, using our work in modeling SDA-enhanced defenses as the exemplar. Second, we discuss lessons learned in developing the appropriate analysis methods to synthesize our results in a form that would yield policy-relevant insights, using the hide/reveal of offensive capabilities as the exemplar. Third, we review lessons learned from our attempts to synthesize and generalize attack strategies. Finally, we discuss lessons learned from our review of the applicability of strategies in chess and go to the space domain. We close with a summary of our experience.

Developing Appropriate Abstractions

We walk through two attempts to properly abstract the defensive advantages of enhanced SDA. Our first abstraction, although it has the virtue of simplicity, yielded a negative result that gave rise to more questions than answers. The second—which is used for the results articulated in the main body of this report—allows us to disentangle different effects of SDA-enhanced defenses but may be obscuring actual entanglement effects. We include this discussion in the hopes that other researchers can learn from our search for an abstraction that is appropriate to our research questions.

Approach 1: Model SDA-Enhanced Defenses as a Reduction in Offensive Effectivity of Kinetic Weapons

For our research, we wanted to explore the defensive advantage of SDA—specifically, its ability to provide warning of a kinetic attack that gives the target an opportunity to mitigate or avoid the impact. Our first attempt at modeling this payoff for improved SDA was simply to reduce the effectiveness of Red's kinetic weapons as a linear function of Blue's possession of SDA capability. This was the simplest abstraction of the effect that we could make, but it turned out to have a number of drawbacks.

Analysis of the results of these games was deeply unsatisfying. We found that our players rarely invested in improving their SDA. If we endowed them with SDA at game start, there appeared to be no deterrent effect on their adversary. As long as the adversary's kinetic weapons had some effectivity, the adversary would use those weapons. In some cases, we did observe attacks on the objects that provided SDA, suggesting that the adversary saw them as a threat. Unfortunately, we were unable to find a pattern to the cases where this occurred.

Upon further reflection, we decided that there were two inadequacies in our modeling that might be obscuring our results. First, our Blue player did not need to take any additional actions to reap the benefits of the SDA investment. In reality, the investment in improved SDA will only have an effect on the success of an adversary's kinetic weapons if a decision is made to act on that intelligence and take defensive actions to mitigate an incoming attack. An additional complication was that, in an attempt to avoid code changes, we had modeled the objects that provided SDA as weapons and suspected the attacks on those objects may have been a result of that modeling choice.

Approach 2: Model SDA-Enhanced Defenses as an Improvement in Defensive Effectivity

To better understand why investments in SDA seemed to have no impact on the game results, we changed our modeling in two significant ways. The first was to model the objects providing SDA as neither assets that provide power projection nor as weapons. Instead, these objects were modeled as purely providing SDA. This meant that any investment in or attack on them would be motivated by the advantage in defense they provided apart from any other objective the players might have. Although this change clarifies cause and effect relationships in using the game for research purposes, in reality, the sensors that provide SDA *are* often integrated into other space objects, making our model less representative of the real world. For this reason, we caution that a lack of attacks on these SDA-providing objects in our games may not be reflective of reality. However, we believe the lack of investment in them is representative of reality: We modeled these investments as relatively low cost, as would be true if they were integrated into other space objects. Hence, the lack of investment is not due to a high cost, but of low return.

Our second change was to model the effectivity of some defensive actions, such as maneuvering or deploying near-field decoys to confuse an incoming kill vehicle, as a function of SDA. If SDA was below a threshold, the defense would be less effective. If SDA was above that threshold, the defense would be more effective. In other words, to reap the benefits of improved SDA, a player would need to *both* invest in SDA and subsequently take defensive action. This, we hypothesized, would allow us to segregate cases in which SDA is of no value because there are other effective deterrents from those in which deterrence has broken down but is restored by the possession of SDA. In the first case, there is no need to take defensive action because the adversary is deterred by something other than our defenses (whether SDA enhanced or not). In the second case, there is no need to take defensive action because the adversary is deterred by the SDA-enhanced effectiveness of the defense. It also let us compare the value of enhanced defenses against the cost of those defenses. This change both clarifies the cause and effect relationships in the model and makes it more representative of reality, in which there is a penalty for taking active defensive action in space.

For instance, maneuvering a satellite requires fuel and reduces overall satellite life. Furthermore, because most satellites are not designed to maintain service performance (e.g., high bandwidth communications or imaging) while maneuvering, defensive maneuvers may also reduce a nation's ability to project power into a distant theater of war for several days. Similarly, decoys can often be deployed only once, making them another non-renewable resource. Furthermore, for decoys to be effective, it is likely that the satellite will need to match its electronic and thermal signature to that of a much simpler decoy, again taking the satellite out of service for a period. Had we not modeled the decision to invest in SDA as separate from the decision to use the enhanced defenses of SDAs, we would not have adequately realized the effects of space being an offensive-dominated warfighting domain in our research.[1]

As this discussion illustrates, finding the right balance between simple models that clarify cause and effect relationships and more complex models that are more representative of reality is not straightforward. Sometimes we can find abstractions that do both, but more often, these two goals are in conflict. This, of course, is not an original thought. George Box perhaps said it best: "All models are wrong, some models are useful."[2]

Developing Appropriate Methods for Analysis and Synthesis of Game Results

Although the previous section illustrated lessons learned about choices in constructing models of reality that are at the appropriate level of abstraction to answer a research question, this section highlights how different analytic approaches affect our ability to conduct insightful research using the game. It uses our work in exploring the effect of misperceptions of offensive capabilities as the exemplar.

[1] Although this second model was deemed adequate for our research purposes, we recognize that it does not capture every potential benefit of SDA. For instance, we made no attempt to model improvements in targeting that could result from better SDA or model what effect that SDA might have on a nation's ability to attribute attacks more convincingly or quickly.

[2] George E. P. Box and Norman R. Draper, *Empirical Model-Building and Response Surfaces*, New York: John Wiley and Sons, 1987.

Analysis Approach 1

Our first approach to assess game outcomes was to look for correlations across a group of variables: Blue investment priorities, Red attack strategy, Red perception of Blue weapon status, and Blue actual weapon status. We compiled game results to observe trends among these variables and develop an intuition about how games are being played. The result of this analysis identified several factors that are correlated with Red aggression, as measured by their willingness to take the first escalatory action in the game. The first is a Blue prioritization of weapons investments, the second is an initial Blue status with a low weapons count, and the third is a Red perception that Blue has a low weapons count. A potential mechanism for these effects is described in Figure C.1.

We then used a linear regression analysis of game outcome metrics and player behavior variables to examine these relationships more quantitatively. This allowed us to explore whether there is a relationship between, for example, initial perceived weapons parity and the tendency toward Red aggression. These trends are extremely noisy, however, and may indi-

FIGURE C.1
Weapon Status and Perceptions Leading to Red Aggression

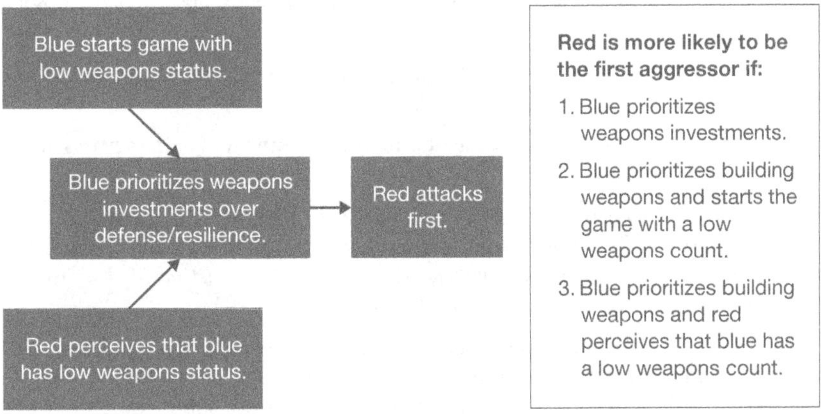

cate that additional factors are required to fully represent the relationship between these variables.[3]

To better clarify why regressions offer little insight for our research, we take a slight detour into the mathematics of regression. Linear regression analysis describes the statistical relationship between an independent variable X (the outcome) and a dependent variable Y (the input).[4] Regression analysis "fits" the coefficients (C) of a linear equation of the input variables, such that $X' = CY$ produces the smallest difference between the observed values of X and the estimated X' that is the output of the equation.[5] In addition to computing the coefficients that denote the strength and direction of the linear relationship between X and Y, the analysis produces a metric for the goodness of the fit overall (termed *R-squared*), along with individual metrics (termed the *p-value*) for each coefficient indicating the statistical significance of the observed correlation.[6]

We now describe a concrete example of what we can and cannot learn from regression analysis. In exploring the effect of misperceptions about an opponent's possession of offensive weapons, we ran the game varying the initial number of offensive weapons, with one player (Blue) starting the game with zero, one, three, or six weapons and the opposing player (Red) starting the game with a *preconceived perception* of the number of weapons Blue has.[7] As discussed in Chapter Three, through other analyses, we found that whether Red overestimated or underestimated the number of weapons

---

3 We observe a low R-squared value, indicating a poor fit, along with a low *p*-value, indicating that there is still a real relationship between these variables despite the poor fit.

4 Or a set of dependent variables $\{Y_1, Y_2, \ldots Y_N\}$ that we hypothesize may be able to predict the outcome X.

5 A common criterion to optimize the fit is to minimize the sum of the squares of the difference between X and X'.

6 The *p*-value for a coefficient indicates whether the contribution of the variable Y to X is distinguishable from simple randomness. For an in-depth discussion of regression methods of analysis, see, for example, David A. Freedman, *Statistical Models: Theory and Practice*, Cambridge, UK: Cambridge University Press, 2009.

7 Games yield a fairly large variation of outcomes that would need to be analyzed with statistical methods because they use the flip of a fair coin to choose between moves that appear to yield identical scores with respect to the player's objectives. This means

that Blue possessed made a policy-relevant significant difference in game outcomes. However, we were unable to determine from that analysis if the degree of over- or underestimation mattered. In an attempt to discover the answer to the latter question, we performed a regression analysis with the primary dependent variables (*Y*) being (1) the quantity of weapons held by Blue at game start and (2) whether Red knows about them. The output variable (*X*) of interest is the conflict intensity of the resulting game. Unfortunately, the regression failed to yield significant insight.

To understand why it failed, we summarize in Table C.1 the output of regression analysis for the two sets of game scenarios, overestimated and underestimated, where we attempt to fit the equation *Conflict Intensity* = *C* * *Initial Quantity of Blue Weapons*. If the initial quantity has an effect on the conflict intensity, then the *p*-value for the coefficient *C* will be below a statistically significant threshold.[8] As seen in the table, for the case where Red overestimates the number of Blue weapons, there is a low *p*-value, indicating that there may be a relationship between the independent variable (conflict intensity) and the dependent variable (number of initial Blue weapons). However, in the case of underestimation, the *p*-value indicates that there is no statistically significant correlation between the magnitude of the underestimation and the resultant conflict intensity. Low R-squared values for both sets of games indicate significant variability in the data—i.e., the equation that even identical initial inputs can result in very different time histories and conflict intensities.

TABLE C.1

Regression Analysis Results Correlating Red Misperception of Quantity of Blue Weapons with Conflict Intensity

	Red Overestimates Blue Weapons Count		Red Underestimates Blue Weapons Count	
	Red	Blue	Red	Blue
R-squared value	0.05	0.05	0.001	0.0003
p-value	0.002	0.002	0.6	0.8

NOTE: Red shading means that there is no statistically significant correlation between the output and input variables of our hypothesis.

[8] Typically, a *p*-value below 0.05 is considered statistically significant.

tion we are trying to fit explains very little about the actual relationships produced by the games.[9]

We could incorporate additional dependent variables to the equation we are trying to fit in an attempt to increase our R-squared metric and perhaps find statistically significant correlations. However, the games we analyze hold all other inputs constant, and we still see low explanatory value for a regression correlating the one input we varied to our output of interest. This means that *other dependent variables must be emergent properties of the game*—i.e., if there is a linear relationship between a set of variables in the game and conflict intensity, those variables are also outputs (not inputs) of the game. This highlights a key limitation in the use of linear regression for the analysis of a complex model, especially one that plays out over time. Although perhaps at any given instant in time we could discover linear relationships between the conditions of and actions in the game, we only can catch a glimpse of relationships between starting conditions and overall game scores. However, as this example illustrates, a regression analysis of our game input/output never produced a simple correlation (i.e., X leads to Y) from which one might formulate policy. It can and did often provide hints about the factors that matter in analyzing game results. In this case, it alerted us to a fundamental difference between under- and overestimation of offensive capabilities.[10]

Analysis Approach 2

To help identify the factors that may be the most impactful on game results, specifically Red deterrence and Blue outcomes, we needed an approach that could provide more clarity about a select variable rather than an approach that simultaneously explores all variables. In short, to build a narrative and better understanding of game play, we needed a hypothesis to test. As described in the main body of this report, at the suggestion of the sponsor,

---

[9] A low R-squared value is not surprising. As discussed earlier, there are literally hundreds of input variables (although kept constant here) and chances in the game. Furthermore, those variables change over time.

[10] In fact, prospect theory—a mindset that we apply to players in some of the games—does produce a bias regarding under- versus overestimation. The regression analysis is picking up on this bias.

we tested to see if our results gave evidence of Sun Tzu's admonition, "when weak appear strong and when strong appear weak."[11] Focusing our analysis on this hypothesis, we compared cases of deception and complete information about the status of Blue's weapons stockpile and found a much more compelling case for the relationship between Red behavior and their perceptions about Blue weapons.

The results of this experiment identify how a nation's ability to project military power from the space domain throughout the time frame of the game is affected by misperceptions. We can see from the results a clear advantage in military power projection when offensive capabilities are hidden, both when the adversary overestimates and underestimates. Interestingly, these results are consistent even when decisionmaking is translated via prospect theory. There is, however, a difference when we change the players' mindsets from a strictly rational approach to one viewed through the lens of prospect theory. Under prospect theory, decisionmaking would lead a player to prefer that the adversary underestimate their offensive capabilities. This is because the adversary underestimates potential gains from an attack and overestimates losses. If weapons are hidden, the adversary also will not perceive risk in maintaining the status quo and will therefore restrain from attacking, leaving the player's assets relatively intact.

Furthermore, this research approach reveals a key challenge in using correlation-based methodologies to interpret AI with a high degree of complexity. Finding correlations and causal relationships that are meaningful may require a sound initial hypothesis to test. Searching for a hypothesis to test was an iterative interaction between human research and AI. Although correlation analysis did help us identify some potential relationships between independent variables and dependent measured outcomes, it could not sufficiently describe how the game outcomes *resulted* from the initial inputs or player models. To test this relationship, we need to build a causal model. Hypothesis development and the construction of counterfactuals are key requirements for setting up an exploration of causation. An important

[11] From Sun Tzu, 1964, "Hence, when able to attack, we must seem unable; When using our forces, we must seem inactive; When we are near, we must make the enemy believe we are far away; When far away, we must make him believe we are near."

insight from our use of the analyses described here was the recognition that building the right hypothesis is vital.

Simplifying the Game: Misadventures with Models

In this section, we discuss experiments that led to ambiguous learning. We document them because (1) we believe other researchers may learn from our work and (2) even failed experiments can yield insight. Throughout the course of this research, the orders of battle used in various games were created to represent realistic future conflict scenarios. Many of these orders of battle were quite asymmetric, and we are reluctant to extract general heuristics about strategy from them. Not only could those asymmetries obscure strategic dynamics, observations about those dynamics might not be extensible beyond those specific conditions. Therefore, after identifying the three strategic interaction patterns using these realistic orders of battle, we set up simpler game runs with symmetric orders of battle. Simpler was not always better.

Brinksmanship: An Early Shaping of the Battlefield Strategy

We hoped to use a symmetric order and then change just one parameter at a time to allow us to develop a more generalizable and causal understanding of the dynamics of space competition and conflict. This did not happen. This did, however, bring to light another mechanism that players can use for early shaping of the battlefield. We term this strategy *brinksmanship*. For many reasons, we cannot rule out the possibility that brinksmanship is a mathematically induced outcome of the game model. Thus, we were hesitant to characterize it as a strategic interaction pattern representative of real-world conflicts. However, in trying to understand it, we generated generalizable learning.

In one set of games that demonstrated the brinksmanship pattern, the asymmetry introduced at game start is that Blue has an asset with a vulnerability that can be exploited using EW attacks. One way to interpret these games is that, in the process of exploiting this vulnerability, Red executes a shaping strategy to effectively tie the hands of Blue by removing nonescalatory response options. The following list describes the general order of actions in the games:

1. Blue invests in hardening vulnerable asset from EW attacks.
2. Red executes EW-reversible attacks against Blue's vulnerable asset during the period from game start until Blue's investment pays off.
3. Blue's investment pays off, making their asset less vulnerable to EW attacks.
4. Red executes kinetic, debris-generating attacks against Blue's asset.
5. Red continues these attacks up until the brink of orbit destruction from debris generation.[12]
6. Neither Blue nor Red can use a kinetic attack unless they are willing to execute a game-ending move.

Because the game starts with a vulnerable Blue asset, Red wants to ensure they exploit this vulnerability in the most effective way. When they see Blue investing in hardening this asset from EW attacks, it creates a use-it-or-lose-it scenario for Red. Red needs to use their EW weapons against this asset before that investment pays out. We are comfortable terming Blue's hardening investment as causal of Red's early use of EW weapons. It is also not difficult to understand why Blue elected to invest in hardening their asset against EW attack. We had initialized the game with this vulnerability, and Blue's move to shore it up is logical, at least in the short term. It is harder to understand why Red moves to the politically costlier debris-generating kinetic attacks once Blue's investment in hardening against EW weapons pays out.

By executing a series of kinetic attacks to degrade Blue's asset, Red brings the game to the brink of orbit destruction. Red creates a scenario in which one more kinetic attack by either player would create enough debris in geosynchronous equatorial orbit to effectively end the game with the loss of the

[12] The potential for debris cascades that result in the loss of use of an orbit is recognized in the scientific literature and is called the *Kessler syndrome*. Although the actual threshold of debris at which these cascades might occur is unknown (or even if there is a threshold), our game models this potential as a known event that occurs with the tenth debris-generating attack in an orbit or, in the case of geosynchronous orbit, in a region of that orbit.

use of that orbit for all parties. By taking the game to the brink of mutual loss of the use of space, Red has taken away Blue's option to use a kinetic attack unless Blue is willing to take on this high cost. In this way, Red has strategically shaped the battlefield to tie Blue's hands in terms of executing kinetic attacks once the ground war starts. Although this accurately describes the effect and perhaps the goal of this strategy, it is more difficult to understand *why* it occurs. We know that moves in the game represent the "best" either player can do while their opponent is simultaneously doing their best. Knowing this set of moves is the "best" either can do in terms of optimizing their objectives, we ask the following questions:

- Why is this brinksmanship strategy in Red's best interest?
- Why is it in Blue's best interest to invest in shoring up their vulnerable assets as opposed to building weapons to use against Red's kinetic weapons? Especially given that Blue has near-perfect knowledge that their investment in hardening against non-kinetic weapons will result in this longer-term dynamic of brinksmanship.[13]
- With orbital destruction at the brink, what would cause either player to push it over?

Although we can make observations about the last question, the answers to the first two are much more difficult to ascertain. It is a known problem with rational game-theoretic models that players will pursue even the slightest advantages to "win" a symmetric game, and the payout of Blue's hardening investment has brought the game back into close symmetry. However, we have mitigated those effects in our game by modeling diminishing returns on attacks (under the assumption that adversaries learn from expe-

[13] Blue does not, however, have perfect knowledge. For computational reasons and to mimic real-world decisionmaking, we limit each player's ability to look ahead to six moves into the future. (Few individuals, including the most skilled chess players, can effectively reason more than a few moves in advance of play.) Therefore, it could be that these games are reflective of that imperfect rationality, *or* it could be that the decision to model imperfect look-aheads change the nature of the game itself. See Vahab Mirrokni, Nithum Thain, and Adrian Vetta, "On the Implications of Lookahead Search in Game Playing," arXiv preprint arXiv:1202.4134, October 31, 2018, for a description of the impact of look ahead search on various game models.

rience) and by stopping the pursuit of advantage below a threshold.[14] Therefore, something else besides the search for minimal gains may be at work in these games of brinkmanship.

The best explanation we can find for why it is in Red's interest to play a game of brinksmanship is that doing so constrains Blue's options, but this is not entirely satisfactory. Mathematically, it means that Red has calculated that Blue can achieve parity in terms of military power projection prior to the start of the ground war and to prevent that from happening, Red attacks early. Therefore, some of Red's attacks are logical. But that still does not explain why Red pushes it to the brink.

Blue makes the initial investment in hardening against EW attack knowing that doing so will result in destruction of the orbit. Mathematically, this means that Blue has calculated that a failure to invest would result in a worse outcome. We believe this is because our scoring rewards maintaining capacity *over time*. From Blue's point of view, it is better to invest and fight Red to the bitter end than to not invest and suffer early defeat. Chess players who value flexibility in response options exhibit this same trait. We believe our math model accurately describes how many nations would think about this issue.

Perhaps it is easiest to understand Red's decision to push the attacks to the brink by thinking about conditions that might cause players to push orbital destruction *over the brink*. After ground war start, each player makes decisions that balance their ability to project military power from space and create vulnerabilities in their opponent's space power projections. If a player's ability to project military power is sufficiently below that of their opponent's, then it is in their best interest to destroy the orbit such that neither side can use it; i.e., each side will be willing to destroy the orbit if the value they gain from continued use of space is less than the value they gain by eliminating the adversary's ability to project power from space. In fact, in the scenario just described, we have observed that if Blue decides to attack Red's assets even via non-kinetic reversible means, such as a cyberattack, it creates asymmetry in the power projection from space that may push Red to

[14] This threshold is set at 0.05 percent of the initial score. If a move will not improve a player's overall score by at least that amount over the course of the game, the move is not played.

execute one last kinetic attack to end the game. Why Blue makes the decision to execute that cyberattack is harder to understand, but Red's response is entirely logical. Blue should only make the cyberattack decision if the orbital destruction cements a small advantage that allows them to "win," and Red should only make the decision to end the game if failure to do so would cause them to "lose more."

This brings us full circle to again ask why Red pushed the game to the brink. Why does the initial setup of this game cause Red to determine that their best option is to construct the conditions so that they can later avoid losing more? Looking across games that exhibit brinksmanship, we found no clear patterns. Furthermore, we found that nearly identical games could end with either Red or Blue executing the final attack that destroyed the orbit. We also observed, as Triezenberg describes in her dissertation (2017), that the addition of prospect theory—where players value the avoidance of loss more than the accumulation of gains—would often transform these same games into games of deterrence. The lesson learned is that parity, or near parity, is stabilizing under prospect theory but destabilizing under rational choice theory. Therefore, understanding your opponent's mindset—how they think about power projection and their rightful place in the world—may be the most critical investment a nation can make in deterring the extension of conflict into outer space.

However, this report is focused on the dynamics of conflict (not deterrence), and, although this scenario prompts us to ask many questions for which we do not have answers, the examination of these games does reveal a key principle that has an analogy in chess: *The dynamic value, both temporally and contextually, of an attack changes with the conditions of the game.* For example, Red's value for using EW attacks is significantly higher prior to Blue's hardening investment pay off. The impact of a Blue cyberattack once the game has been pushed to the brink of destruction is much greater than if that same attack had been executed earlier in the game. Finally, the impact of the tenth kinetic attack (the one that generates the critical debris level to effectively destroy the orbit) is much higher than the first kinetic attack.

Although we failed to achieve our goal of producing a more generalizable understanding of our strategic interaction patterns via a simple symmetric game, this exercise did in fact enhance our understanding of generalizable principles of conflict. Failure in this case—and even the possible use of a flawed mathematical model—proved essential to learning.

Impact of Investments on Strategic Interaction Patterns

In our search for understanding the conditions that determine which of the three strategic interaction patterns are used, we also conducted a set of analyses to examine if there was a correlation between Blue's initial investment decisions and Red's attack decisions. Specifically, we wanted to understand whether Blue's initial decision to invest in redundancy and resilience of assets rather than offensive capabilities (i.e., weapons) changed the strategic interaction patterns. We quickly learned two things from this analysis:

1. We had no way to segregate Blue's initial investment decisions from the initial conditions of the game. The resulting game dynamics can be attributed as much to the initial game setup as to Blue's initial investment decisions.
2. Attack patterns are most easily explained by *recognizing the dynamic nature of the value of different weapon types*. This was determined by the introduction and removal of different types of hardening from the bundle of moves.

As just discussed, when we set up a symmetric order-of-battle game with a small Blue vulnerability, Blue initially invests in redundancy and resilience to shore up that vulnerability (in the case above, hardening against EW attacks). Recognizing the diminishing effectiveness of their weapons if they wait, Red responds by engaging in attacks early in the game to exploit Blue's vulnerability before those attacks become significantly less effective (a use-it-or-lose-it play). However, although this is one condition that leads to Red strategies of *early shaping of the battlefield*, we do not believe that this use-it-or-lose-it condition explains all instances in which we found this strategy. In other, more complex games, we observed that the early shaping could be equally well explained as an attempt by the opponent to force investments in redundancy and resilience at the expense of building weapons. In fact, in several of those games, we suspected that it was an initial investment in weapons that spurred the early attacks.

Therefore, we also set up a symmetric order-of-battle game with a small Red asymmetric weapons advantage. This had the expected effect of switching Blue's initial investments to building weapons in an attempt to achieve weapons parity. Red's response in this case is to prevent these investments from paying out by engaging in less destructive but distracting reversible attacks. This is still an attempt to shape the battlefield early in the game, but instead of engaging in attacks to exploit Blue's diminishing vulnerability, Red instead is seeking to avoid creating their own vulnerability to new Blue weapons. If Red decides to escalate to kinetic attacks in these games, this escalation takes place at ground war start in a war of attrition or via horizontal escalation to space after ground war start. Red recognizes that the use of kinetic weapons before ground war start has an associated political cost that may not be worth it if Blue has sufficient defenses or resilience to mitigate the impact of an attack. If Red waits until ground war start to engage in kinetic attacks, it will not incur this political cost. Furthermore, attacks on assets prior to ground war start incurs political penalty for no military gain, but attacks on assets after ground war start provides military gain to offset the political penalties. Unless Red is in a use-it-or-lose-it scenario, they decide to wait to avoid paying the political cost. This reflects the dynamic value of weapons in the game and illustrates how players can maximize the value of weapons use (or minimize the cost of weapons use). If a Red weapon will lose value if not used early in the game, they will not wait to use this weapon. If a Red weapon will incur cost if used early, but not if used later in the game, they will wait to use it. This recognition of the dynamic value of weapons is a key insight in our study of these game results.

Impact of Concealment on Strategic Interaction Patterns

> The history of warfare from antiquity to the present records innumerable attempts to secure by some new contrivance an immediate tactical advantage, perhaps a decisive one. In such inventions the essential

> purpose is to obtain one's end before the adversary can bring countermeasures to bear. It is the time interval that counts.[15]

As this quote suggests, one way to maximize an investment in a new technology is to shorten the time interval for an opponent to build countering technologies. A time-honored means of doing so is to hide the possession of a technology for as long as possible. Although playing the technology would reveal its existence, it may be possible to hide it until that moment. Therefore, we gave our players the ability to hide their initial allocations of both assets, weapons, and SDA. We also gave them the ability to reveal or deceive whether they had those capabilities.

Although we were able to learn a few things from the ability to hide, we only learned the limitations of our game-theoretic model by giving our players the ability to reveal and deceive. To mimic reality, we made the payout for a reveal dependent on a 50-percent probability that a player would believe the reveal and revise their estimates of their opponent's capabilities closer to truth and an equal probability that they would view the reveal as a bluff and either not revise their estimate or revise it further away from truth. We invoked similar but opposite rules for deception. Because players can observe the payouts of moves in game-theoretic models, we quickly learned that if the payout moved the opponent in the direction advantageous to the player, the player would continue to reveal/deceive. If, however, the payout moved in the direction opposite of the desired effect, the player would abandon attempts to reveal/deceive. As a result, these moves were rarely used, and any attempt to find a pattern was obscured by the randomness of the payouts.[16]

Another unfruitful investigation was the effect of hiding initial allocations of SDA. As we detailed in Chapter Three, possession of SDA largely had no impact on game outcomes and, by extension, neither did the abil-

---

[15] From Bernard Brodie, *Sea Power in the Machine Age*, Princeton, N.J.: Princeton University Press, 1941.

[16] We may have learned a great deal about the dynamics of hide/reveal had we run sets of games that varied the probabilities of belief versus bluff. In fact, if we had infinite resources (or even faster computers than the supercomputers used), we undoubtedly would have pursued such an investigation. Instead, we have elected to revert to simpler game trees in forthcoming research to explore these issues further.

ity to hide possession of SDA. The ability to hide weapons, however, did lead to the interesting results outlined in Chapter Three—a clear confirmation of Sun Tzu's precepts of the advantages of deception of weakness and strength. After having discovered the strategic interaction patterns, we returned to these games to search for correlations between what was hidden and those patterns. Unfortunately, at the time of the writing of this report, we have found no consistent correlations. However, our research into the value of revealing and concealing assets is ongoing in the form of investigating smaller Bayesian signaling games.

Comparisons to Chess and Go

We conclude this appendix with a discussion of how the patterns of strategic interactions we found in our game of space war are or are not similar to those found in two of the most popular games of strategy: chess and go. We find that we can use the games of chess and go to help us illustrate the three conflict strategies we observed in our games: *early shaping of the battlefield*, *wars of weapons attrition*, and *horizontal escalation*.

The best analogue in chess for the *early shaping of the battlefield* are openings: established series of moves that are considered optimal in the first few moves in a game of chess. These openings generally focus on fulfilling a few intermediate objectives of chess. Of particular interest to our discussion are the objectives of "developing pieces" and "controlling the center." Pieces in chess are considered *developed* when they are moved off their respective backline. These are important intermediate objectives because of how pieces derive their value in a game of chess. The value of a piece at any given point of a game of chess is determined by how many squares it threatens (could move to) and its ability to threaten squares in the future. Because of this, moving pieces off of the backline and into the center, or into a position that threatens the center, increases their value. In a similar vein, players also want to prevent or dissuade an opponent from developing their own pieces and controlling the center. This is where the connection to a shaping-the-battlefield strategy is made. Shaping the battlefield is characterized by making specific investments, similar to developing your pieces, that force your opponent to respond in ways that do not necessarily help them achieve their intermediate goals (i.e., developing their own pieces or controlling the center). With these general ideas in mind, we now look at different broad styles of openings in chess.

There are many approaches to building control of the center in the early stages of a game, but two of the most popular are the classic and hypermodern styles of play. The classic method for controlling the center of the board is through direct occupation of the center positions, either by using pawns or other pieces. In contrast, the hypermodern method uses pieces on the flanks, establishing control through influence and not occupation. For example, a player adopting this method will often position their bishops to threaten squares in the center board.

The transition from the early game to the middle game is difficult to describe largely because the middle game does not have a particularly precise definition. For this discussion, let us say that the middle game is simply the phase of the game that is not the early game, when most of your pieces are undeveloped, and it is not the endgame, when most of your pieces are captured. We know a few things about the middle game, primarily that it involves a large number of developed pieces threatening a large number of squares and the frequent capture of pieces during the transition to the endgame. This second part aligns well with the strategy of *wars of weapons attrition*. In chess, the general heuristic when trading pieces is that you should trade major pieces (non-pawns) when you have an advantage and trade pawns when you are behind. The reason for this general heuristic is that when you are ahead, it tends to be beneficial to reduce your opponent's options to regain their initiative, whereas when you are behind, trading pawns reduces the future potential of both players. Both of these tactics have the same underlying idea: constraining your opponent's decision space through a conflict of attrition. This plays out in other aspects of the middle game, which we will describe as defensive and offensive play. *Defensive play* seeks to constrain moves sets for both players, finding ways to manipulate the opponent down another path and reduce the number of options available to either side. Players may pin the opponent pieces in a way that allows them to maintain control of their movement but deny them any benefit from doing so. By taking away these gains, a player has effectively deterred an opponent action. In contrast, *offensive play* seeks to maximize one's own decision space, regardless of what the opponent may do in response.

The endgame of chess provides some analogue to the horizontal strategy found in our space competition games through situations in which a player has pawns on opposite sides of the board advancing, but this is not a particularly apt illustration. For a stronger example of *horizontal escalation*, we turn to the game of go.

A significant difference between chess and go is in their objectives. Chess has the sole objective of catching the king. Although there are many methods to achieve this objective, in the end, the only piece that matters is the king. Go is an imperialistic game where a player's goal is to enclose more territory on the board than their opponent. Furthermore, chess is hierarchical in that different pieces have different inherent value. Each of the pieces in go has equal inherent power, and their value depends on how they are played. Also unlike chess, which resembles a more tactical fight of maneuvering limited pieces, go is focused on territory control with immobile pieces. This is not to say that pieces in go are static, rather they change with the overall landscape of the board instead of moving themselves. In chess, the pieces of the game are moved around the board, and this can actively change their value. However, in go, the pieces do not move around the board, and their value is controlled by how other pieces are positioned. This is exacerbated by the fact that go boards are much larger, 19 × 19 squares rather than 8x8, allowing for several different distinct conflicts to happen at the same time with the potential to spill over into each other. This is where go play intersects with the horizontal escalation strategy in our games. In go, multiple battles can exist simultaneously, but ultimately these different battles will affect each other either directly or indirectly. Direct spillover is fairly obvious, and the two conflicts get close enough to each other to the point where pieces placed in one battle affect the outcome of another and vice versa. Indirect spillover is more abstract and is based on the concept of initiative. In sequential games, players have to choose what limited move they will make before their opponent can act, and, therefore, multiple conflicts can draw your opponent's moves away from one conflict and toward another. Both direct spillover and forcing your opponent to allocate resources elsewhere are broadly analogous to the horizontal escalation strategy in our games. The concept of forcing your opponent to allocate resources elsewhere is also, of course, an attribute of an early shaping of the battlefield strategy:

- Sacrifices and exchanges: Use short-term, tactical actions to shape the board for long-term advantage.
- Immediate profit: Simultaneous objectives of materiel profit/territory and mobility/influence.

The two objectives played in chess—to manipulate the opponent and to maximize one's own decision space—are also observed in our space competition game play. A classic player is valuing power projection through early control of the board, analogous to the *early shaping of the battlefield* approach we observed in our games. A hypermodern player is valuing flexibility in move and response options, moving pieces into a position that allows them to better mobilize and respond to adversary actions. This player behavior is similar to our players adding moves to their move set (see Appendix B for more detailed description of game play), analogous to the development of TTPs in military operations, with the idea that these moves may never get used but instead serve to make it easier to deploy these moves if needed. In a *war of weapons attrition*, we also observe player behavior analogous to the aggressive chess player in the middle stages of the game. For horizontal escalation, we can find parallels in go as battles in one area of the board spill over into others either directly or indirectly.

Summary of Using Artificial Intelligence to Conduct Research

In this report, we offer only tentative observations regarding the conditions that give rise to the observed strategic interaction patterns and possible ways a nation might shape an opponent's attack strategies. Although we can re-create the logic for any individual decision made by the players in our games, clear correlations that might lead us to causal theories for why a particular strategy emerges are less readily observable.

In fact, our experience in using complex AI to explore strategy leads us to offer three observations about that use:

1. We have come to believe that a complex game-theoretic model is incapable of revealing anything beyond correlation and, even then,

only if we have a theoretical basis for our inquiry. A hypothesis can then be used to construct counterfactual game scenarios that can be run to test whether A does in fact lead to B.[17] More often than not, however, we found that factors we hypothesized might be causal had no impact on overall strategies. There are simply too many substitutes available in the game for any one factor to be causal.[18]

2. We find ample evidence of the butterfly effect of chaos theory in our work. Very small differences in initial conditions or individual decisions, which we might term negligible in the short term, often have profound effects in the longer term. As Edward Lorenz states in his seminal talk on the butterfly effect, the best model for research may not be the most exact model of the world.[19] Simpler models may be more useful.[20]
3. None of the research conducted with the model proceeded in a linear fashion. Runs with our initial models inevitably taught us some key aspect of the problem. We then taught the model to incorporate that

[17] This approach is called *causal AI*. For a top-level view, see Judea Pearl and Dana Mackenzie, *The Book of Why: The New Science of Cause and Effect*, New York: Basic Books, 2018.

[18] Where we could construct and find evidence to support specific theories, that support depended on *combinations* of factors. However, we were unsuccessful in our attempts to use pattern-recognition algorithms, such as those used in machine learning, to detect combinations of factors that would be explanatory of strategies. We suspect this is because we did not have adequate instrumentation to capture emergent properties over the course of the game.

[19] The text of Lorenz's December 1972 speech can be found in Ralph Abraham and Yoshisuke Ueda, eds., *The Chaos Avant-Garde: Memoirs of the Early Days of Chaos Theory*, World Scientific Series on Nonlinear Science, Series A, Vol. 39, Singapore: World Scientific, 2000.

[20] Slantchev, 2017, argues the particular case that simpler game-theoretic models are most useful in the study of conflict. However, as he notes, the decision of what is "simple enough" is not straightforward. This appendix discusses some of our more-notable failures in modeling appropriate abstractions in the hope that others may benefit of our experience.

learning, sometimes through multiple iterations, with the net result being what we term *incremental human-AI collaborative learning*.[21]

The following discussion provides, we hope, some perspective on the three observations. The second observation about the butterfly effect arose from our game-changer analysis. Although the game-changer analysis reliably indicates significant changes in game outcomes, in many cases, we found only barely distinguishable differences in the progression of game conditions that led to those significantly different outcomes. Eventually, we learned that even games with *identical* initial conditions and move sets could result in very different outcomes. This we attribute to the small set of places in the game where we flip a fair coin to determine the action a player will take when multiple courses of action appear to yield identical results. Although any individual decision may be indistinguishable from another at the time a decision is made, it can have far-reaching results when played out over the ten years, providing ample evidence of the butterfly effect of chaos theory in our work.

If we accept the presence of chaos in our games, it leads to important observations about how we can use complex games to develop strategic insight. Although re-creating the logic used for strategic decisionmaking in the game has yielded key insights about player behavior and rationale, chaos theory provides a clear caution against using complex AI to *predict* strategic decisionmaking. Instead, an AI such as ours should be used to help identify *causal* relationships between player decisionmaking and game outcomes. If we understand why, we are far more empowered to develop a better strategy. And if we do *not* understand the causes of behavior, either ours or our

[21] A reasonable body of applicable research in human-AI collaborative learning comes from observations about how the game of chess has changed since the first computers became smart enough to defeat the leading human strategists. For a top-level view, see Dirk Knemeyer and Jonathan Follet, "How 22 Years of AI Superiority Changed Chess," *Towards Data Science*, March 5, 2019. A more academic treatise on the subject of human-AI collaborative learning can be found in Nan-ning Zheng, Zi-yu Liu, Peng-ju Ren, Yong-qiang Ma, Shi-tao Chen, Si-yu Yu, Jian-ru Xue, Ba-dong Chen, and Fei-yue Wang, "Hybrid-Augmented Intelligence: Collaboration and Cognition," *Frontiers of Information Technology and Electronic Engineering*, Vol. 18, No. 2, 2017.

adversary's, we could easily make ineffective or even quite poor decisions. But, as we have seen, identifying causal relationships is hard.

If we approach the task of finding causal factors using probabilistic associations between variables, this is essentially just finding regularities in data sets too large to discern with our naked eye or "little more than curve-fitting" as asserted by Judea Pearl, a pioneer in the field of causality and AI. She goes on to argue that truly intelligent machines would reason using causation, not probabilistic association, and would ask counterfactual questions within this causal framework.[22] Ultimately, the goal of AI is to be able to infer the causal relationships from the data entirely. But if the model is too complex, we cannot build this inference. The few correlations we found proved limiting in that they did not allow us to go further and determine how the game outcomes resulted from the initial inputs or player models. Recall, for example, that denying our players the ability to make game-changing moves that were highly correlated with outcomes had no impact on outcomes. For this reason, we look to *causal* AI, in contrast to *predictive* AI, as a guide to how AI tools can be used to develop strategic insight.

Causal AI helps identify precise relationships between cause and effect while also facilitating the modeling of interventions and their impact on outcomes. Although predictive models build understanding about what might happen, causal models build understanding of the levers at your disposal to change these outcomes. Understanding these levers is at the heart of our research using this game over the years and highlights that causal AI is a more apt approach for these types of research questions.

Hypothesis development and the construction of counterfactual are key requirements for setting up an exploration of causation. In this, causal AI techniques mirror the randomized controlled trial that is the gold standard for exploring causal effects. A randomized controlled trial approach splits the subjects (in our case, games) into a control group and a group subject to the intervention being explored. It was only by constructing and running games that denied our players the ability to make game-changing moves (the counterfactual) that were we able to demonstrate that those moves were not, in fact, game changing. From this, we were then able to derive the prin-

[22] Pearl, 2018.

ciple that the diversity of weapon/target pairings in space means that adding better defenses against a specific class of weapons is unlikely to change an opponent's strategy in future space wars.[23]

However, setting up an randomized controlled trial requires an understanding of which intervention we should explore. Not all counterfactuals are as easy to set up and explore as the elimination of a move from the game. In the case of our explorations about the effect of possessing and perceptions of possession of co-orbital kinetic weapons discussed in Chapter Three, we needed to build a counterfactual scenario. This need required that we answer the question "what distinguishes the scenarios?" Is the distinction whether Blue has weapons initially, or is it whether Red *perceives* that Blue has weapons? This distinction defines the hypothesis we are testing, and without this initial hypothesis, identifying causal relationships becomes impossible. In fact, in this case, it was only after we had refined the hypothesis one step further—whether it was the perception of weakness or of strength (as opposed to possession)—that we were able to offer a tentative causal statement.

Effective hypotheses emerged *as a result of a human-AI collaborative process.* Initial game runs would yield possible correlations, prompting us to build hypotheses and then adjust the game to explore further hypotheses in an incremental human-AI collaborative learning process. A build, learn, rebuild, and learn-again cycle was essential to our work. By recounting this experience, we hope others will be better prepared when conducting their own exploratory research using complex AI. For us, reflecting on this experience has driven a return to much simpler game-tree models in the hopes of discovering hypotheses that we can then bring to our analysis of these more complex games. We believe it is essential to concentrate on hypothesis formation if we are to make further progress in understanding the conditions that give rise to the strategies of space war observed in the game.

[23] We are confident in asserting that we would not have discovered this principle had we not constructed the counterfactual.

Abbreviations

AI	artificial intelligence
EW	electronic warfare
GPS	Global Positioning System
GSSAP	Geosynchronous Space Situational Awareness Program
M	military
P	political
PLA	People's Liberation Army
PMSII	political, military, social, information, and infrastructure
SDA	space domain awareness
SSF	Strategic Support Force
SSI	social, information, and infrastructure
TDRSS	Tracking and Data Relay Satellite System
TTP	tactics, techniques, and procedures
USSF	U.S. Space Force

References

Abraham, Ralph, and Yoshisuke Ueda, eds., *The Chaos Avant-Garde: Memoirs of the Early Days of Chaos Theory*, World Scientific Series on Nonlinear Science, Series A, Vol. 39, Singapore: World Scientific, 2000.

Air Force Space Command (Archived), "Geosynchronous Space Situational Awareness Program," webpage, March 22, 2017a. As of May 25, 2021: https://www.afspc.af.mil/About-Us/Fact-Sheets/Article/730802/geosynchronous-space-situational-awareness-program-gssap/

Air Force Space Command (Archived), "Space Based Space Surveillance," webpage, March 22, 2017b. As of May 25, 2021: https://www.afspc.af.mil/About-Us/Fact-Sheets/Article/249017/space-based-space-surveillance-sbss/

Allison, Graham, *Destined for War: Can America and China Escape Thucydides Trap?* Boston: Houghton Mifflin Harcourt, 2017.

Axelrod, Robert, *The Evolution of Cooperation*, New York: Basic Books, 1984.

Bigelow, David, "An Analysis of the Richardson Arms Race Model," November 25, 2003. As of May 25, 2021: http://jfrabajante.weebly.com/uploads/1/1/5/5/11551779/arms_race_models.pdf

Bodner, Matthew, "Russian Military Merges Air Force and Space Command," *Moscow Times*, August 3, 2015. As of May 25, 2021: https://www.themoscowtimes.com/2015/08/03/russian-military-merges-air-force-and-space-command-a48710

Bowen, Bleddyn E., *War in Space: Strategy, Spacepower, Geopolitics*, Edinburgh: Edinburgh University Press, 2020.

Box, George E. P., and Norman R. Draper, *Empirical Model-Building and Response Surfaces*, New York: John Wiley and Sons, 1987.

Brodie, Bernard, *Sea Power in the Machine Age*, Princeton, N.J.: Princeton University Press, 1941.

Brodie, Bernard, *The Anatomy of Deterrence*, Santa Monica, Calif.: RAND Corporation, RM-2218, 1958. As of May 25, 2021: https://www.rand.org/pubs/research_memoranda/RM2218.html

CNA Center for Autonomy and AI, "Impact of Unmanned Systems to Escalation Dynamics," undated. As of January 2019: https://www.cna.org/CNA_files/PDF/Summary-Impact-of-Unmanned-Systems-to-Escalation-Dynamics.pdf

Debreu, Gerard, *Theory of Value: An Axiomatic Analysis of Economic Equilibrium*, New Haven, Conn.: Yale University Press, 1959.

Department of Defense Instruction 4650.08, *Positioning, Navigation, and Timing and Navigation Warfare*, Washington, D.C.: U.S. Department of Defense, December 27, 2018, change 1 effective December 30, 2020. As of June 10, 2021:
https://www.esd.whs.mil/Portals/54/Documents/DD/issuances/dodi/465008p.pdf?ver=M9B6zSt5uWSeDoPwocp_RQ%3D%3D

Duggan, William, *Strategic Intuition, The Creative Spark in Human Achievement*, New York: Columbia Business School Publishing, [2007] 2013.

Erwin, Sandra, "Air Force: SSA Is No More, It's 'Space Domain Awareness'," *SpaceNews*, November 14, 2019. As of May 25, 2021:
https://spacenews.com/air-force-ssa-is-no-more-its-space-domain-awareness/

Finch, James P., and Shawn Steene, "Finding Space in Deterrence: Toward a General Framework for 'Space Deterrence'," *Strategic Studies Quarterly*, Vol. 5, No. 4, Winter 2011, pp. 10–17.

Freedman, David A., *Statistical Models: Theory and Practice*, Cambridge, UK: Cambridge University Press, 2009.

Goirigolzarri, Benjamin, *A Need for Speed? Identifying the Effects of Space Acquisition Timelines on Space Deterrence and Conflict Outcomes*, Santa Monica, Calif.: RAND Corporation, RGSD-432, 2019. As of May 12, 2020:
https://www.rand.org/pubs/rgs_dissertations/RGSD432.html

Gosnold, "Contested Space II: Countermeasures," SatelliteObservation.net, March 8, 2018. As of May 25, 2021:
https://satelliteobservation.net/2018/03/08/contested-space-ii-countermeasures/

Halpern, Joseph Y., "Computer Science and Game Theory: A Brief Survey," in S. N. Durlauf and L. E. Blume, eds., *Palgrave Dictionary of Economics*, May 2007. As of November 2020:
https://www.cs.cornell.edu/home/halpern/papers/csgt.pdf

Harrison, Todd, Kaitlyn Johnson, Thomas G. Roberts, Tyler Way, and Makena Young, *Space Threat Assessment 2020: A Report of the CSIS Aerospace Security Project*, Washington, D.C.: Center for Strategic and International Studies, 2020. As of November 2020:
https://csis-website-prod.s3.amazonaws.com/s3fs-public/publication/200330_SpaceThreatAssessment20_WEB_FINAL1.pdf

Hitchens, Theresa, "Exclusive: NRO, SPACECOM Craft CONOPS for War in Space," *Breaking Defense*, May 4, 2020. As of May 25, 2021:
https://breakingdefense.com/2020/05/exclusive-nro-spacecom-craft-conops-for-war-in-space/

Joint Publication 3-0, *Joint Operations*, Washington, D.C.: Joint Chiefs of Staff, January 17, 2017, incorporating change 1, October 22, 2018.

Kahneman, Daniel, and Amos Tversky, "Prospect Theory: An Analysis of Decision Under Risk," *Econometrica,* Vol. 47, No. 2, March 1979, pp. 263–292.

Kessler, Donald J., and Burton Cour-Palais, "Collision Frequency of Artificial Satellites: The Creation of a Debris Belt," *Journal of Geophysical Research*, Vol. 83, No. A6, June 1, 1978, pp. 2637–2646.

Klein, John J., *Space Warfare: Strategy, Principles and Policy*, New York: Routledge, 2006.

Kleinberg, Howard, "On War in Space," *Astropolitics: The International Journal of Space Politics and Policy*, Vol. 5, No. 1, 2007, pp. 1–17.

Knemeyer, Dirk, and Jonathan Follet, "How 22 Years of AI Superiority Changed Chess," *Towards Data Science*, March 5, 2019. As of January 2019: https://towardsdatascience.com/how-22-years-of-ai-superiority-changed-chess-76eddd061cb0?gi=ceed27132f42

Los Angeles Air Force Base, "Section 31: Bringing the Space and Missile Systems Center's Software Factory to Life," webpage, September 3, 2019. As of May 25, 2021: https://www.losangeles.spaceforce.mil/News/Article-Display/Article/1952209/section-31-bringing-the-space-and-missile-systems-centers-software-factory-to-l/

Merriam-Webster, "bluff," dictionary entry, undated-a. As of May 25, 2021: https://www.merriam-webster.com/dictionary/bluff

Merriam-Webster, "feint," dictionary entry, undated-b. As of May 25, 2021: https://www.merriam-webster.com/dictionary/feint

Merriam-Webster, "opportunity cost," dictionary entry, undated-c. As of May 25, 2021: https://www.merriam-webster.com/dictionary/opportunity%20cost

Merriam-Webster, "tactics," dictionary entry, undated-d. As of May 25, 2021: https://www.merriam-webster.com/dictionary/tactics

Mirrokni, Vahab, Nithum Thain, and Adrian Vetta, "On the Implications of Lookahead Search in Game Playing," arXiv preprint arXiv:1202.4134, October 31, 2018.

Morgan, Forrest E. *Deterrence and First-Strike Stability in Space: A Preliminary Assessment*, Santa Monica, Calif.: RAND Corporation, MG-916-AF, 2010. As of May 25, 2021: https://www.rand.org/pubs/monographs/MG916.html

Myerson, Roger B., *Game Theory: Analysis of Conflict*, Cambridge, Mass.: Harvard University Press, 1991.

Pearl, Judea, and Dana Mackenzie, *The Book of Why: The New Science of Cause and Effect*, New York: Basic Books, 2018.

Pollpeter, Kevin L., Michael S. Chase, and Eric Heginbotham, *The Creation of the PLA Strategic Space Force and Its Implications for Chinese Military Space Operations*, Santa Monica, Calif.: RAND Corporation, RR-2058-AF, 2017. As of May 25, 2021:
https://www.rand.org/pubs/research_reports/RR2058.html

Rapoport, Anatol, "Lewis F. Richardson's Mathematical Theory of War," *Journal of Conflict Resolution*, Vol. 1, No. 3, September 1957.

Reesman, Rebecca, and James R. Wilson, *The Physics of Space War: How Orbital Dynamics Constrain Space to Space Engagements*, El Segundo, Calif.: Aerospace Corporation, October 2020. As of November 2020:
https://aerospace.org/sites/default/files/2020-10/Reesman_PhysicsWarSpace_20201001.pdf

Richardson, Lewis F., *Statistics of Deadly Quarrels*, Quincy Wright and C. C. Lienau, eds., Pittsburgh, Pa.: Boxwood Press, 1960.

Richardson, Lewis F., Nicholas Rashevsky, and Ernesto Trucco, eds., *Arms and Insecurity: A Mathematical Study of the Causes and Origins of War*, Pittsburgh, Pa.: Boxwood Press, 1960.

Rubinstein, Ariel, "Comments on the Interpretation of Game Theory," *Econometrica*, Vol. 59, No. 4, July 1991, pp. 909–924.

Schelling, Thomas C., "The Diplomacy of Violence," in John Garnett, ed., *Theories of Peace and Security*, London: Palgrave Macmillan,1970, pp. 64–84.

Slantchev, Branislav L., "On the Proper Use of Game-Theoretic Models in Conflict Studies," *Peace Economics, Peace Science and Public Policy*, Vol. 23, No. 4, December 14, 2017.

Space Force News, "Next X-37B Orbital Test Vehicle Scheduled to Launch," webpage, May 6, 2020. As of May 25, 2021:
https://www.spaceforce.mil/News/Article/2177702/next-x-37b-orbital-test-vehicle-scheduled-to-launch/

Sun Tzu, *The Art of War*, trans. Samuel B. Griffith, New York: Oxford University Press, 1964.

Torrington, Geoffrey, Bonnie L. Triezenberg, Krista Langeland, Lisa Saum-Manning, Timothy Marler, Elizabeth M. Bartels, and James Pita, *Exploring Space Deterrence: Final Phase II Report: Using Game Theory and Prospect Theory to Inform Future Strategies*, Santa Monica, Calif.: RAND Corporation, 2019, Not available to the general public.

Triezenberg, Bonnie L., *Deterring Space War: An Exploratory Analysis Incorporating Prospect Theory into a Game Theoretic Model of Space Warfare*, Santa Monica, Calif.: RAND Corporation, RGSD-400, 2017. As of May 12, 2020:
https://www.rand.org/pubs/rgs_dissertations/RGSD400.html

U.S. Department of Defense, "Department of Defense Establishes U.S. Space Force," press release, December 20, 2019. As of May 25, 2021:
https://www.defense.gov/Newsroom/Releases/Release/Article/2045981/department-of-defense-establishes-us-space-force/

U.S. Space Force, *Space Capstone Publication: Spacepower, Doctrine for Space Forces*, Arlington, Va.: Headquarters, U.S. Space Force, June 2020. As of May 25, 2021:
https://www.spaceforce.mil/Portals/1/Space%20Capstone%20Publication_10%20Aug%202020.pdf

Von Neumann, John, and Oskar Morgenstern, *Theory of Games and Economic Behavior*, Princeton, N.J.: Princeton University Press, 1944.

Weedon, Brian, and Victoria Samson, *Global Counterspace Capabilities: An Open Source Assessment*, Broomfield, Colo.: Secure World Foundation, April 2020. As of November 2020:
https://swfound.org/media/206957/swf_global_counterspace_april2020_es.pdf

Williams, J. D., *The Compleat Strategyst: Being a Primer on the Theory of Games and Strategy*, Santa Monica, Calif.: RAND Corporation, [1954] 2007. As of May 25, 2021:
https://www.rand.org/pubs/commercial_books/CB113-1.html

Zheng, Nan-ning, Zi-yi Liu, Peng-ju Ren, Yong-qiang Ma, Shi-tao Chen, Si-yu Yu, Jian-ru Xue, Ba-dong Chen, and Fei-yue Wang, "Hybrid-Augmented Intelligence: Collaboration and Cognition," *Frontiers of Information Technology and Electronic Engineering*, Vol. 18, No. 2, 2017, pp. 153–159.

CPSIA information can be obtained
at www.ICGtesting.com
Printed in the USA
LVHW060720170723
751165LV00035B/313

9 781977 408099